DOT

The Duality of Time Postulate and Its Consequences on General Relativity and Quantum Mechanics

Mohamed Haj Yousef

PAPERBACK: ISBN-13: 978-1687895509
All rights reserved. No part of this book may be reprinted or reproduced or utilized in any form or by any electronic, mechanical, or other means, now known or hereafter invented, including photocopying and recording, or in any information storage or retrieval system, without permission in writing from the author.
Copyright © Mohamed Haj Yousef
UNITED ARAB EMIRATES UNIVERSITY
First Release: September 2019

This book is extracted from the previous three volumes in the Single Monad Model of the Cosmos series, published between 2014 and 2019:

2014 - The Single Monad Model of the Cosmos (Ibn Arabi's Concept of Time and Creation) - Paperback: ISBN-13: 978-1499779844, ISBN-10: 1499779844

2017 - Duality of Time (Complex-Time Geometry and Perpetual Creation of Space) - Paperback: ISBN-13: 978-1987778250, ISBN-10: 1987778251

2019 - Ultimate Symmetry (Fractal Complex-Time, the Incorporeal World and Quantum Gravity) - Paperback: ISBN-13: 978-1723828690, ISBN-10: 1723828696

More information on the websites:
http://www.smonad.com
http://www.singlemonad.com
http://www.ibnalarabi.com

Dedication:

... to the Greatest Master Sheikh Muhyiddin Ibn al-Arabi and all his students.

Contents

Dedication 3
Preface 9
Acknowledgment 12
        The SMONAD.COM Website. 13

The Duality of Time Postulate and Its Consequences on General Relativity and Quantum Mechanics 15
- 1- Abstract: 15
- 2- Introduction: 15
- 3- The Duality of Time Postulate: 18
- 4- A Brief General Analysis: 18
- 5- The Dynamic Formation of Dimensions: 23
  - 5.1- The Genuinely-Complex Time-Time Frame: 25
  - 5.2- The Two Arrows of Time: 28
  - 5.3- General Relativity Approximation: 32
  - 5.4- Aether and the Cosmological Constant Problem: 36
  - 5.5- Origin of Mass 39
- 6- Deriving the Principles of Special and General Relativity: 40
  - 6.1- Lorentz Transformations: 43
  - 6.2- The Mass-Energy Equivalence Relation: 45
  - 6.3- The Equivalence Principle of General Relativity 57
  - 6.4- Complex Energy 59
- 7- Fractal Space-Time and Quantum Phenomena: 61
  - 7.1- Super-symmetry and its Breaking: 64
  - 7.2- The Exclusion Principle: 65
  - 7.3- Uncertainty: 65
  - 7.4- Collapse of Wave-Function: 66
  - 7.5- Schroedinger's Cat: 67
  - 7.6- Entanglement and Non-locality: 67
  - 7.7- Causality: 68
- 8- Conclusion. 69
- 1- Appendix: Deducing $E = mc^2$ from $m = \gamma m_0$: 70

Author Biography 71

References and Bibliography 84

List of Figures

1 Representing the inner and outer levels of time 24
2 The actual velocity and the complex velocity 45
3 Gradual versus abrupt change of speed 49

Introduction

This short book presents a brief and concise exploration of the Duality of Time postulate and its consequences on General Relativity and Quantum Mechanics. To make it easier for citing, this book is presented in the form of a scientific paper, which will also make it more accessible and easier to be read by researchers who are interested in the astounding conclusions rather than any exhausting introductions which are provided in the previous books for more general readability.

The purpose of this book is to demonstrate how the Duality of Time Theory, that results from the Single Monad Model of the Cosmos, could explain and solve many major problems in physics and cosmology, including causality, non-locality, homogeneity, the arrow of time, the mass-gap (and Yang-Mills Conjecture), super-symmetry and matter-antimatter asymmetry, in addition to uniting the principles of Relativity and Quantum theories, as well as the psychical and spiritual domains; all based on the same genuinely-complex time-time geometry that provides the smooth link between the corpuscular physical particles and the absolute homogeneous (Euclidean) space, via four distinctive levels of symmetry: normal, super, hyper and ultimate, in addition to the original level of absolute Oneness.

All these symmetries will be naturally incorporated in the Standard Model of Elementary Particles, after discovering the hidden granular geometry revealed by the duality of time theory that explains how multiplicity is dynamically emerging from absolute Oneness, at every instance of our normal time! Therefore, this theory leads to the Ultimate Symmetry of space and its instantaneous breaking into the physical and psychical creations, with two hyperbolically orthogonal arrows of time.

The **Single Monad Model of the Cosmos** book series contains three volumes:

The Single Monad Model of the
Cosmos: Ibn al-Arabi's Concept
of Time and Creation
ISBN-13: 978-1499779844
ISBN-10: 1499779844
Publication Date: June, 2014

Duality of Time Theory:
Complex-Time Geometry and
Perpetual Creation of Space
ISBN-13: 978-1499779844
ISBN-10: 1499779844
Publication Date: December, 2017

Ultimate Symmetry: Fractal
Complex-Time, the Incorporeal
World and Quantum Gravity
ISBN-13: 978-1723828690
ISBN-10: 1723828696
Publication Date: January, 2019

In the first volume, we introduced Ibn al-Arabi's eccentric conception of time and outlined the general aspects of his cosmological views. This fundamental insight was developed in the second volume into the Duality of Time Theory, which provided elegant solutions to many persisting problems in physics and cosmology, including the arrow-of-time, super-symmetry, matter-antimatter asymmetry, wave-particle duality, mass generation, homogeneity, and non-locality, in addition to deriving the principles of Special and General Relativity based on its granular complex-time geometry. The third volume explores how the apparent physical and metaphysical multiplicity is emerging from the oneness of divine presence, descending through four fundamental levels of symmetry: ultimate, hyper, super and normal, in addition to the original level of absolute oneness.

The Duality of Time does not contradict the established theories of Modern Physics, but it exposes a deeper understanding of time that leads to the metaphysical (non-background) complex-time space, which is absolutely Euclidean, and which approximates to the non-Euclidean, Minkowskian or de Sitter / anti- de Sitter, space when we suppose the background to be continuously existing with real physical dimensions. This approximation leads to General Relativity, and the current first and second quantization, of energy and fields, but Quantum Gravity requires this third quantization, that is the true quantization of the space-time itself, and this cannot be achieved without taking into account the hidden symmetry of the new granular complex-time geometry.

Therefore, the correspondence principle is fulfilled by the Duality of Time Theory, because the semi-Riemannian geometry on the real space R is a special approximation of this complex-time geometry on the hyperbolic space H, so General Relativity is produced when we consider space and matter to be coexisting together in (and with) time, thus causing the deceptive continuity of physical existence. All the principles of Special and General Relativity can be derived mathematically from the Duality of Time postulate, in addition to exact derivation of the mass-energy equivalence relation that is not possible within the current theories without introducing some approximation or induction.

This Duality of Time Theory does not only explain how the Universe appeared *ex-nihilo*, but that it is always in this ontological state of *potential existence*; only perpetually coming into being for some series of consecutive discrete instances of time, the duration of which is absolutely zero, but as a result of our manipulative past memory and future expectation we imagine time extensions in which we are no more than dreamlike passing shadows making some monotonous noise for a short period of *imaginary* time, before we leave this World to a relatively higher level of existence.

Therefore, this research on the Single Monad Model made a substantial breakthrough in mathematics, physics, and cosmology, as well as natural philosophy, because it exposes a deeper level of time and original theory of creation that could explain many persisting problems in these various fields. The main result of the Duality of Time Theory is that *vacuum* is the real time, while our time is genuinely imaginary or latent to it, which makes it described as *void*. This vacuum-void duality is basically the same ancient atomistic philosophy, reinterpreted out of the most original theological and philosophical conceptions, combined together with the conclusions of modern physics and cosmology.

This research took more than three decades to crystallize, and it may be several more decades are needed to realize its immense consequences, because it goes far beyond the long-awaited Theory-of-Everything, to include all the physical, psychical and spiritual domains. This unique understanding of geometry will cause a paradigm shift in our knowledge of the fundamental nature of the cosmos and its corporeal and incorporeal structures. There is no doubt that this is the most significant discovery in the history of mathematics, physics and philosophy, ever!

The SMONAD.COM Website

The website: http://www.smonad.com is dedicated for the Single Monad Model of the Cosmos, including this second book and the previous one, and it will contain related articles and extracts in addition to readers contributions and a forum for comments and other discussions on the subject of time.

Acknowledgment

I would like to express my truthful gratitude to my master Sheikh Ramadan Subhi Deeb, of Sheikh Ahmad Kuftaro Foundation in Damascus, for his continuous support and inspirational insights that he always offered to me throughout the course of this long research over the past two decades. My sincere appreciation and thanks are also due to Prof. James W. Morris, of Boston College, who supervised me in the University of Exeter during the Ph.D. project that lead to the Single Monad Model which produced the Duality of Time Theory.

Author:

Mohamed bin Ali Haj Yousef
Sunday, September 1, 2019
UAE University, Al-Ain
United Arab Emirates

The Duality of Time Postulate and Its Consequences on General Relativity and Quantum Mechanics

1 Abstract:

Based on the Single Monad Model and Duality-of-Time hypothesis, a dynamic and self-contained space-time is introduced and investigated. It is shown that the resulting "time-time" geometry is genuinely complex, fractal and granular, and that the non-Euclidean space-time continuum is the first global approximation of this complex-time space in which the (complex) momentum and energy become invariant between different inertial and non-inertial frames alike. Therefore, in addition to Lorentz transformation, the equivalence principle is derived directly from the new discrete symmetry. It is argued that according to this postulate, all the principles of relativity and quantum theories can be derived and become complementary. The Single Monad Model provides a profound understanding of time as a complex-scalar quantum field that is necessary and sufficient to obtain exact mathematical derivation of the mass-energy equivalence relation, in addition to solving many persisting problems in physics and cosmology, including the arrow-of-time, super-symmetry, matter-antimatter asymmetry, mass generation, homogeneity, and non-locality problems. It will be also shown that the resulting physical vacuum is a perfect super-fluid that can account for dark matter and dark energy, and diminish the cosmological constant discrepancy by at least 117 orders of magnitude.

2 Introduction:

Relativity, and its classical predecessor, consider space and time to be continuous, and everywhere differentiable, whereas quantum mechanics is based on discrete quanta of energy and fields, albeit they still evolve in continuous background. Although both theories have already passed many rigorous tests, they inevitably produce enormous contradictions when applied together in the same domain. Most scholars believe that this conflict may only be resolved with a successful theory of quantum gravity (Calcagni, 2017).

In trying to resolve the discrepancy, some space-time theories, such as Causal Dynamical Triangulation (Ambjørn et al., 2005), Quantum Einstein Gravity (Lauscher and Reuter, 2005) and Scale Relativity (NOTTALE, 1992), attempted to relax the condition of differentiability, in order to allow for fractal space-time, which was first introduced in 1983 (Ord, 1983). In addition to the abundance of all kinds of fractal structures in nature, this concept was also supported by many astronomical observations which show that the Universe exhibit a fractal aspect over a fairly wide range of scales (Joyce et al., 2005), and that large-scale structures are much better described by a scale-dependent fractal dimension (Hogg et al., 2005), but the theoretical implications of these observations are not very well understood, yet.

Nonetheless, the two most celebrated approaches to reconcile Relativity with Quantum Mechanics are Strings Theory and Loop Quantum Gravity (LQG). The first tries to develop an effective quantum field theory of gravity at low energies, by postulating strings instead of point particles, while LQG uses spin networks to obtain granular space that evolves with time. Therefore, while Strings Theory still depends on the background continuum, LQG tries to be background-independent by attempting to quantize space-time itself (Rovelli, 2011).

In this regard, the author believes that any successful theory of quantum gravity must not rely on either the continuum or discretuum structures of space-time. Rather, these two contrasting and mutually-exclusive views must be the product of such theory, and they must become complementary on the microscopic and macroscopic scales. The only contestant that may fulfill this criterion is "Oneness", because on the multiplicity level things can only be either discrete or continuous; there is no other way. However, we need first to explain how the apparent physical multiplicity can proceed from this metaphysical oneness, and then exhibit various discrete and continuous impressions. The key to resolve this dilemma is in understanding the "inner levels of time" in which "space" and "matter" are perpetually being "re-created" and layered into the three spatial dimensions, which then kinetically evolve throughout the "outer level of time" that we encounter. This will be fully explained in sections 3 and 5 below.

Due to this "dynamic formation of dimensions", in the inner levels of time, the Duality of Time Theory leads to **granular** and **self-contained** space-time with **fractal** and **genuinely-complex** structure, which are the key features needed to accommodate both quantum and relativistic phenomena. Many previous studies have already shown how the principles of quantum mechanics can be derived from the fractal structure of space-time (Nottale and Célérier, 2007; JUMARIE, 2001; Cresson, 2003; Adda and Cresson, 2005; Jumarie, 2007), but they either do not justify the use of fractals, or they are forced to make new unjustified assertions, such as the relativity of scale, that may lead to fractal space-time. On the other hand, imaginary time had been successfully used in the early formulation of Special Relativity by Poincare (Poincaré, 1906), and even Minkowski (Einstein, 2010), but it was later replaced by the Minkowskian four-dimensional space-time, because there were no substantial reasons to treat time as imaginary. Nevertheless, this concept is still essential in current cosmology and quantum field theories, since it is employed by Feynman's path integral formulation, and it is the only way to avoid singularities which are unavoidable in General Relativity.

In the Duality of Time Theory, since the dimensions of space and matter are being re-created in the inner (complete) levels of time, the final dimension becomes multi-fractal and equals to the dynamic ratio of "inner" to "outer" times. Additionally, and for the same reason, space-time becomes "genuinely complex", since both its "real" and "imaginary" components have the same nature of time, which itself becomes as simple as the "recurrence", or counting the number of geometrical nodes as they are re-created in one chronological sequence. Without postulating the inner levels of time, both the complex and fractal dimensions would not have any "genuine" meaning, unless both the numerator and denominator of the fraction, and both the real and imaginary parts of the complex number, are all of the same nature (of time).

In this manner, normal time is an imaginary and fractional dimension of the complete dimensions of space, which are the real levels of time. Because they are complete integers, the dimensions of space are mutually perpendicular, or spherically orthogonal, on each other, which is what makes (isotropic and homogeneous) Euclidean geometry that can be expressed with normal complex numbers $\mathbb{C}$, in which the modulus is given by $|z| = \sqrt{x^2 + y^2}$. In contrast, because it is fractional or non-integer dimension, (normal, or the outer level of) time is hyperbolically orthogonal on the dimensions of space, and thus expressed by the hyperbolic split-complex numbers $\mathbb{H}$, in which the modulus is given by $\|z\| = \sqrt{x^2 - y^2}$. This complex hyperbolic geometry is the fundamental reason behind relativity and Lorentz transformations, and it provides the required tools to express the curvature and topology of space-time, away from Riemannian manifolds, in which the geometry becomes ill-defined at the points of singularities.

3 The Duality of Time Postulate:

The Duality of Time Theory, and the resulting dynamic re-creation of space and matter, is based on previous research that presented an eccentric conception of time (Haj Yousef, 2005, 2007, 2014, 2018, 2017, 2019, which include other references on the history and philosophical origins of this concept). For the purpose of this article, this hypothesis can be extracted here into the following postulate:

- At every instance of the outward normal level of time, space and matter are perpetually being re-created in one chronological sequence, which forms the inner levels of time that are also nested inside each lower dimension of space.

4 A Brief General Analysis:

The above postulate means that at every instance of the "real flow of time" there is only one metaphysical point, that is the unit of space-time geometry, and the Universe is a result of its perpetual recurrence in the "inner levels of time", that is continuously re-creating the dimensions of space and what it may contain of matter particles, which then kinetically evolve throughout the outer (normal) level of time, that we encounter.

To understand this complex flow of time, we need to define at least two frames of reference. The first is our normal *3D* "space"

4 A Brief General Analysis: 19

container which evolves in the outer time, that is the normal time that we encounter. And the second frame is the inner flow of time, that is creating the dimensions of space and matter. This inner frame is also composed of more inner levels to account for the creation of $2D$ and $1D$ space, but we shall not discuss them at this point.

From our point to view, as observers situated in the **outer frame**, the creation process is **instantaneous**, because we only see the Universe after it is created, and we don't see it in the **inner frames** when it is being created, or perpetually re-created, at every instance. Nevertheless, the speed of creation, in the innermost level (or real flow) of time, is **indefinite**, rather than **infinite**, because there is nothing to compare to it at this level of absolute oneness. We shall show that this speed of creation is the same speed of light, and the reason why individual observers, situated in the outer frame, measure a **finite** value of it is because they are subject to the **time lag** during which the spatial dimensions are being re-created.

Therefore, in our outer frame, the speed of creation, that is the speed of light, is simply equal to the ratio of the outer to inner times, so it is a unit-less number whose normalized value corresponds to the fractal dimension of the genuinely-complex time-time geometry, rather than space-time, since space itself is created in the inner levels of time. The reason why this cosmological speed is independent of the observer is because creation is occurring in the inner real levels while physical motion is in the outer (normal) time that is flowing in the orthogonal dimension with relation to the real dimensions of space (or inner levels of time).

In other words, while the real time is flowing unilaterally in one continuous sequence, creating only one metaphysical point at every instance, individual observers witness only the discrete moments in which they are re-created, and during which they observe the dimensions of space and physical matter that have just been re-created in these particular instances; thus they only observe the collective (physical) evolution as the moments of their own time flows by, and that's why it becomes imaginary, or latent, with relation to the original real flow of time that is creating space and matter.

Therefore, the speed of light in complete vacuum is the speed of its dynamic formation, and it is undefined in its own reference frame (as it can be also inferred from the current understanding of time dilation and space contraction of special relativity), because the physical dimensions are not yet defined at this metaphysical level. Observers in all other frames, when they are re-created, measure a finite value of this speed because they have to wait their turn until the re-creation process is completed, so any minimum action, or unitary motion, they can do is always delayed by an amount of time proportional to the dimensions of vacuum (and its matter contents if they are measuring in any other medium). Hence, this maximum speed that they can measure is also invariant because it is independent of any physical (imaginary) velocity, since their motion is occurring in the outer time dimension that is orthogonal onto the spatial dimensions which are being re-created in the inner (real) flow of time.

This also means that all physical properties, including mass, energy, velocity, acceleration and even the dimensions of space, are emergent properties; observable only on the outward level of time, as a result of the temporal coupling between at least two geometrical points or complex-time instances. Moreover, just like the complete dimensions of space, the outer time itself, which is a fractional dimension, is also emerging from the same real flow of time that is the perpetual recurrence of the original geometrical point. This metaphysical entity that is performing this creation is called "the Single Monad", that has more profound characteristics which we don't need to analyze in this paper (see (Haj Yousef, 2014, Ch. VI) for more details); so we only consider it as a simple abstract or dimensionless point: $0D$.

It will be shown in section 6 how this single postulate leads at the same time to all the three principles of Special and General Relativity together, since there is no more any difference between inertial and non-inertial frames, because the instantaneous velocity in the imaginary time is always "zero", whether the object is accelerating or not! This also means that both momentum and energy will be "complex" and "invariant" between all frames, as we shall discuss further in sections 6.3 and 6.4 below.

Henceforth, this genuinely-complex time, or time-time geometry will define a profound discrete symmetry that allows expressing the (deceitfully continuous) non-Euclidean space-time in terms of its granular and fractal complex-time space, whose granularity and fractality are expressed through the intrinsic properties of hyperbolic numbers ($\mathbb{H}$), i.e. without invoking Riemannian geometry, as discussed further in section 5.1. However, this hidden discrete symmetry is revealed only when we realize the internal chronological re-creation of spatial dimensions; otherwise if we suppose their continuous existence, space-time will still be hyperbolic but not discrete. Discreteness is introduced when the internal re-creation is interrupted to manifest in the outward normal time, because creation is processed sequentially by the perpetual recurrence of one metaphysical point, so the resulting complex-time is flowing either inwardly to create space, or outwardly as the normal time, and not both together.

Therefore, in accordance with the correspondence principle, we will see in section 5.3, that semi-Riemannian geometry on $\mathbb{R}^4$ is a special approximation of this discrete complex-time geometry on $\mathbb{H}^4$. This approximation is implicitly applied when we consider space and matter to be coexisting together in (and with) time, thus causing the deceptive continuity of physical existence, which is then best expressed by the non-Euclidean Minkowskian space-time continuum of General Relativity, or de Sitter/anti-de Sitter space, depending on the value of cosmological constant.

For the same reason, because we ideally consider the dimensions of space to be continuously existing, our observations become time-symmetric, since we can apparently-equally move in opposite directions. Therefore, this erroneous time-symmetry is reflected in most physics laws because they also do not realize the sequential metaphysical re-creation of space, and that is why they fail in various sensitive situations such as the second law of Thermodynamics (the entropic arrow of time), Charge-Parity violations in certain weak interactions, as well as the irreversible collapse of wave-function (or the quantum arrow of time).

In the Duality of Time Theory, the autonomous progression of the real flow of time provides a straightforward explanation of this outstanding historical problem. This will be explicitly expressed by equation 1, as discussed further in section 5.2, where we will also see that we can distinguish between three conclusive states for the flow of complex-time: either the imaginary time is larger than the real time, or the opposite, or they are equal. Each of the first two states forms one-directional arrow of time, which then become orthogonal, while the third state forms a two-directional dimension of space, that can be formed by or broken into the orthogonal time directions.

This fundamental insight could provide an elegant solution to the problems of super-symmetry and matter-antimatter asymmetry at the same time, as we shall discuss in section 5.2.

Additionally, the genuine complex-time flow will be employed in section 6.2 to derive the mass-energy equivalence relation $E = mc^2$, in its simple and relativistic forms, directly from the principles of Classical Mechanics. This should provide a conclusive evidence to the Duality of Time hypothesis, because it will be shown that an exact derivation of this experimentally verified relation is not possible without considering the inner levels of time, since it incorporates motion at the speed of light which leads to infinities on the physical level. All current derivations of this critical relation suffer from unjustified assumptions or approximations (Einstein, 1905; Planck, 1906; Ives, 1952; Einstein, 1907; Ohanian, 2009), as was also repeatedly acknowledged by Einstein himself (Einstein, 1907; Capria, 2005).

Finally, as an additional support to the Duality of Time Theory, we will show in section 5.4 that the resulting dynamic quintessence will diminish the cosmological constant discrepancy by at least 117 orders of magnitude. This huge difference results simply from realizing that the modes of quantum oscillations of vacuum occur in chronological sequence, and not all at the same time. Therefore, we must divide by the number of modes included in the unit volume, to take the average, rather than the collective summation as it is currently treated in quantum field theories. The remaining small discrepancy could be also settled based on the new structure of physical vacuum, which is shown to be a perfect super-fluid. The Duality of Time Theory, therefore, brings back the same classical concept of aether but in a novel manner that does not require it to affect the speed of light, because it is now the background space itself, being granular and re-created dynamically in time, and not something in a fixed background continuum that used to be called vacuum. On the contrary, this dynamical aether provides a simple ontological reason for the constancy and invariance of the speed of light, which is so far considered an axiom that has not been yet proven in any theoretical sense.

The Duality of Time Theory provides a deeper understanding of time as a fundamental complex-scalar quantum field that reveals the discrete symmetry of space-time geometry. This revolutionary concept will have tremendous implications on the foundations of physics, philosophy and mathematics, including geometry and number theory; because complex numbers are now genuinely natural, while the reals are one of their extreme, or unrealistic, approximations. Many major problems in physics and cosmology can be resolved according to the Duality of Time Theory, but it would be too distracting to discuss all that in this introductory article. The homogeneity problem, for example, will instantly cease, since the Universe, no matter how large it could be, is re-created sequentially in the inner levels of time, so all the states are synchronized before they appear as one instance in the normal level. Philosophically also, since space-time is now dynamic and self-contained, causality itself becomes a consequence of the sequential metaphysical creation, and hence the fundamental laws of conservation are simply a consequence of the Universe being a closed system. This will also explain non-local and non-temporal causal effects, without breaking the speed of light limit, in addition to other critical quantum mechanical issues, some of which are outlined in other publications (Haj Yousef, 2014, 2017, 2019).

5 The Dynamic Formation of Dimensions:

According to the above Duality of Time postulate, the dynamic Universe is the succession of instantaneous discrete frames of space, that extend in the outward level of time that we normally encounter, but each frame is internally created in one chronological sequence within each inward level of the real flow of time. This is schematically demonstrated in Figure 1, where space is conventionally shown in two dimensions, as the (x,y) plane, and we will mostly consider the x axis only, for simplicity.

In reality, however, we can conceive of at least **seven levels of time**, which *curl* to make the four dimensions of space-time: $3D+T$, that are the three spatial and one temporal dimensions; since each spatial dimension is formed by two of the six inner levels, as we shall explain further in section 5.2, while the seventh is the outer time that we normally encounter.

As it will be explained further in section 5.2 below, each spatial dimension is dynamically formed by the real flow of time, and whenever this flow is interrupted, a new dimension starts, which is achieved by multiplying with the imaginary unit that produces

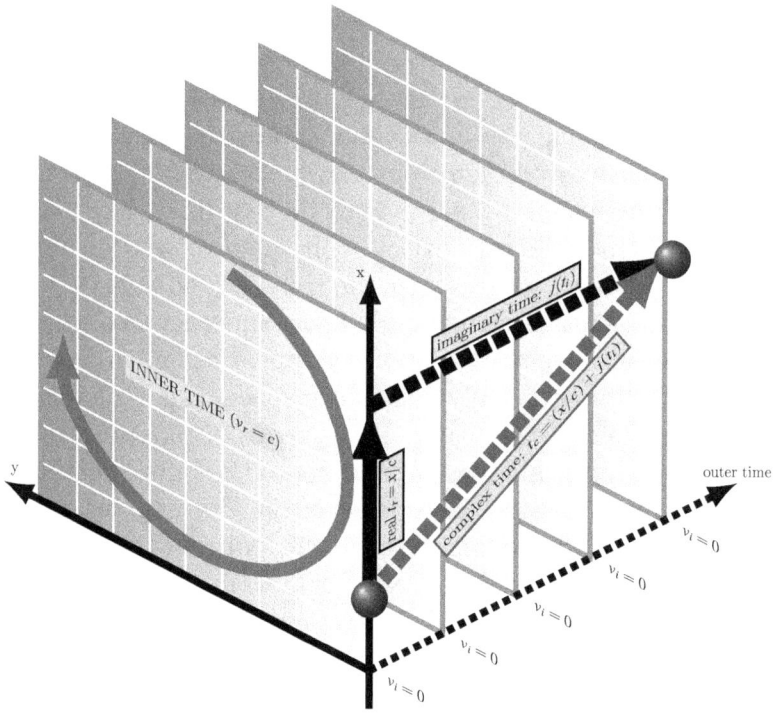

Figure 1: Representing the inner and outer levels of time together as genuinely-complex time-time space. In the real flow of time, one frame of space is re-created in internal chronological sequence, that then appears as one instance of the imaginary time t_i, so the total complex time is: $t_c = (x/c) + j(t_i)$, as split-complex or hyperbolic numbers. We can also notice that the instantaneous velocity v_i in the normal time is always zero, while the real speed v_r in the inner time is always c, and the apparent physical velocity v is the dynamic combination between them as can be calculated from equation 5.

an "abrupt rotation" by $\pi/2$, creating a new dimension that is perpendicular on the previous level, or hyperbolically orthogonal on it, to be more precise. This subtle property is what introduces discreteness, as a consequence of the duality nature of time, that is flowing either inwardly or outwardly, not both together. This is what makes space-time geometry genuinely complex and granular, otherwise if we consider all the dimensions to be coexisting together it will appear continuous and real, as we normally "imagine", which may lead to space-time singularities at extreme conditions.

The concept of imaginary time is already being used widely in various mathematical formulations in quantum physics and cosmology, without any actual justification apart from the fact that it is a quite convenient mathematical trick that is useful in solving many problems. As Hawking states: "It turns out that a mathematical model involving imaginary time predicts not only effects we have already observed, but also effects we have not been able to measure yet nevertheless believe in for other reasons." (Hawking, 1998).

Hawking, however, considers the imaginary time as something that is perpendicular to normal time that exists together with space, and that's how it is usually treated in physics and cosmology. According to the Duality of Time postulate, however, since space is now (dynamically re-created in) the real time, the normal time itself becomes genuinely imaginary, or latent.

Employing imaginary time is very useful because it provides a method for connecting quantum mechanics with statistical mechanics by using a Wick rotation, by $\pi/2$. In this manner we can find a solution to dynamics problems in n dimensions, by transposing their descriptions in $n+1$ dimensions, i.e. by trading one dimension of space for one dimension of time, which means substituting a mathematical problem in Minkowski space-time into a related problem in Euclidean space. Schroedinger equation and the heat equation are also related by Wick rotation. This method is also used in Feynman's path integral formulation, which was extended in 1966 by DeWitt into gauge invariant functional-integral (DeWitt, 1967). For this reason, there has been many attempts to describe quantum gravity in terms of Euclidean geometry (Hawking, 1979; Panine and Kempf, 2016), because in this way it is possible to avoid singularities which are unavoidable in General Relativity, since it is primarily constructed on curved space-time continuum that uses Riemannian manifolds, in which the geometry becomes ill-defined at the points of singularities.

5.1 The Genuinely-Complex Time-Time Frame:

Mathematically, the nested levels of time can be represented by imaginary or complex numbers where space is treated as a plane or spherical wave, and time is the orthogonal imaginary axis. However, in addition to the normal complex number plane: $\mathbb{C}$, that can describe Euclidean space, split-complex, or hyperbolic, numbers: $\mathbb{H}$, are required to express the relation between space-like and time-like dimensions, which are the inner and outer levels of time, respectively. Normal complex numbers can describe homogeneous or isomorphic space, without (the outer) time, where each number z defines a circle, or sphere, because its modulus is given by $|z| = \sqrt{x^2+y^2}$, while in split-complex numbers the modulus is given by $\|z\| = \sqrt{x^2-y^2}$, so $\|t_c\| = \sqrt{t_r^2 - t_i^2}$, which defines a hyperbola. This negative sign in calculating the modulus of complex time t_c reflects the essential fact that the perpetual re-creation of space and matter particles in the inner levels of time t_r is interrupted and re-newed every instance of the outward time t_i, which produces kinetic motions on the physical level, as dynamic local deformations of the otherwise flat and homogeneous Euclidean space.

Therefore the non-Euclidean Minkowski space-time coordinates (x,y,z,t) are an approximation of the complex space-time coordinates (x,y,z,jct_i), or complex time-time coordinates: $(x/c, y/c, z/c, jt_i)$ where $(x/c, y/c, z/c)$ or (t_x, t_y, t_z) represent the inner **real part of time**: $t_r = \sqrt{t_x^2 + t_y^2 + t_z^2} = \sqrt{(x/c)^2 + (y/c)^2 + (z/c)^2}$, and (t_i) represents the outer **imaginary time**, so the total **complex time** is $t_c = t_r + jt_i$, where j is the split-imaginary unit which defines hyperbolic numbers, thus: $t_c \in \mathbb{H} \equiv \mathbb{R}(j)$, and its modulus is $\|t_c\| = \sqrt{t_r^2 - t_i^2}$. These hyperbolic numbers have been also called: tessarines, motors, bireal, perplex, semi-complex, or split-complex, but in this article, unless otherwise stated, *imaginary* and *complex* refer to these hyperbolic numbers $\in \mathbb{H}$ (Rochon and Shapiro, 2004).

In this abstract complex frame, space and time are absolute, or mathematical, just as they had been originally treated in the classical Newtonian Mechanics, but now empty space is **void**, which is a pure mathematical space, because it does not have any material reality, to differentiate it from the physical **vacuum**, which is the dynamic aether, or the ground state of matter. So, for void both the real and imaginary parts are null: $(0,0)$, while for vacuum only the imaginary part is null: $(t_r, 0)$, which indicates infinite and inert space that is the ground state of matter particles, that are then described by (t_r, t_i), or: (c, v), which means that they are internally being re-created at the speed of c, that is the real part, and moving outwardly at the apparent velocity v, that is the imaginary part;

given by equation 5.

The physical vacuum, which is the dynamic aether, is therefore an extreme state which may be achieved when the apparent velocity, or momentum, becomes absolutely zero, both as the object's total velocity and any vector velocities of its constituents, and this corresponds to absolute zero temperature ($0K$). This dynamic vacuum state is therefore a super-fluid, which is a perfect Bose-Einstein condensate (BEC), since it consists of indistinguishable geometrical points that all share the same state. In quantum field theory, complex-scalar fields are employed to describe superconductivity and superfluidity (Lan, 1965). The Higgs field itself is complex-scalar, and it is the only fundamental scalar quantum field that has been observed in nature, but there are other effective field theories that describe various physical phenomena. Indeed, some cosmological models have already suggested that vacuum could be a kind of yet-unknown super-fluid, which would explain all the four fundamental interactions and provide mass generation mechanism that replaces or alters the Higgs mechanism that only partially solves the problem of mass. In BEC models, masses of elementary particles can arise as a result of interaction with the super-fluid vacuum, similarly to the gap generation mechanism in superconductors (Zloshchastiev, 2011), in addition to other anticipated exotic properties that could explain many problems in the current models, including dark matter and dark energy (HUANG, 2013; HUANG et al., 2012). Therefore, the new complex-time geometry is the natural complex-scalar quantum field that explains the dynamic generation of space, mass and energy. We will discuss the origin of mass in sections 5.5 and 6.2.3 below.

Actually, according to this genuinely-complex time-time geometry, there can be four absolute or "super" states: super-mass $(0,0)$, super-fluid $(c,0)$, super-gas $(0,c)$, and super-energy (c,c), which can be compared with the classical four elements: earth, water, air, and fire, respectively. These four extreme or elemental states, which the ancient Sumerians employed in their cosmology to explain the complexity of Nature, are formed dynamically, in the inner levels of time, by the Single Monad that is their "quint-essence". We will see, in section 5.4 below, that this new concept of aether and quintessence is essential for understanding dark matter and energy, and solving the cosmological constant discrepancy.

Moreover, the super-fluid and super-gas states, $(c,0)$ and $(0,c)$, are in orthogonal time directions, so if (c,v) describes matter that is kinetically evolving in the normal level of time with v velocity, (v,c) would similarly describe anti-matter in the orthogonal direction. This could at once solve the problems of super-symmetry and matter-antimatter asymmetry, because fermions in one time direction are bosons in the orthogonal dimension, and vice versa, and of course these two dimensions do not naturally interact because they are mutually orthogonal. This could also provide some handy tests to verify the Duality of Time Theory, but this requires prolonged discussion beyond the scope of this article, as outlined in other literature (Haj Yousef, 2017, 2019). Super-symmetry and its breaking will be also discussed further in section 7.1.

5.2 The Two Arrows of Time:

Discreteness implies interruption or discontinuity, and this is what the outer time is doing to the continuous flow of the inner time that is perpetually re-creating space and matter in one chronological sequence. Mathematically, this is achieved by multiplying with the imaginary unit, which produces an "abrupt rotation" by $\pi/2$, creating a new dimension that is orthogonal on the previous level. Multiplying with the imaginary unit again causes time to become real again, i.e. like space. This means that each point of our $3D+1$ space-time is the combination of seven dimensions of time, the first six are the real levels which make the three spatial dimensions, $t_r = \sqrt{(x/c)^2 + (y/c)^2 + (z/c)^2}$, and the seventh is the imaginary level that is the outer time t_i.

This outward (normal) level of time, t_i, is interrupting and delaying the real flow of time, t_r, so it can not exceed it, because they both belong to one single existence that is flowing either in the inward levels to form the continuous (real) spatial dimensions, or in the outward level to form the imaginary discrete time, not the two together; otherwise they both would be real as we are normally deceived. As we introduced in section 3 above, the reason for this deception is because we only observe the physical dimensions, in the outer time, after they are created in the inner time, so we "imagine" them to be co-existing continuously, when in fact they are being sequentially re-created. It is not possible otherwise to obtain self-contained and granular space-time, whose geometry could be defined without any

5 The Dynamic Formation of Dimensions:

previous background topology. Thus, we can write:

$$0 \leq t_i \leq t_r. \tag{1}$$

So because t_i is interrupting and delaying t_r, the actual (net value of) time is always smaller than the real time: $\|t_c\| = \sqrt{t_r^2 - t_i^2} \leq t_r$, and this is actually the proper time as we shall see in equation 3. However, it should be noted here that, unlike the case for normal complex (Euclidean) plane, the modulus of split-complex numbers is different from the norm, because it is not positive-definite, but it has a metric signature $(1, -1)$. This means that, although our normal time is flowing only in one direction because it is interrupting the real flow of creation and can not exceed it, it is still possible to have the orthogonal state where the imaginary time is flowing at the speed of creation and the real part is interrupting it, such that: $t_c = t_i + jt_r$, so: $\|t_c\| = \sqrt{t_i^2 - t_r^2}$, and then $t_r \leq t_i$, from our perspective. In this case, the ground state of that vacuum would be $(0, c)$, which describes anti-matter as we shall explain further in section 7.1, when we speak about super-symmetry and its breaking.

Equivalently, the apparent velocity v can not exceed c because it is the average of all instantaneous velocities of all individual geometrical points that constitute the object, which are always fluctuating between 0 and c; so by definition v is capped by c, as expressed by equation 5.

Therefore, equation 1 ($0 \leq t_i \leq t_r$) is also equivalent to: $0 \leq |v| \leq c$, thus when: $t_i \longrightarrow t_r = x/c$, we get: $v = x/t_i \longrightarrow c$, and if $t_i = 0$, then $v = 0$; but both as the total apparent velocity of the object and any vector velocities of its constituents, thus in this case we have flat and infinite Euclidean space without any motion or disturbance, which is the state of vacuum: $(t_r, 0)$, or $(c, 0)$, as we noted in section 5.1. So this imaginary time, t_i, is acting like a resistance against the perpetual re-creation of space, and its interruption, i.e. going in the outward level of time, is what causes physical motion and the inertial mass m_0, which then effectively increases with the imaginary velocity according to: $m = \gamma m_0$ (as we shall derive it in section 6.2, and we shall discuss mass generation in sections 5.5 and 6.2.3), and when the outward imaginary time approaches the inner time, the apparent velocity v approaches the speed of creation c, and $m \longrightarrow \infty$. If this extreme state could ever happen (but not by acceleration, as we shall see further below), the system would be described by (c,c), which means that both the real and imaginary parts of complex-time would be continuous, and this describes another homogeneous Euclidean space with one higher dimension than the original $(c,0)$ vacuum.

Actually, the hyperbolic split-complex number (c,c) is non-invertible null vector that describes the asymptotes, whose modulus equals zero, since both its real and imaginary parts are equal. At the same time, as a normal complex number, (c,c) describes an isotropic infinite and inert Euclidean space (without time), because its dimensions are continuous, or uninterrupted. The metaphysical entities of the Universe are sequentially oscillating between the two vacuum states (as Euclidean spaces or normal complex numbers): $(c,0)$ and (c,c), while collectively they appear to be evolving according to the physical (hyperbolic) space-time states (c,v), as split-complex numbers. Therefore, the vacuum state can be described either as the non-invertible vector (c,c) in the hyperbolic plane $\mathbb{H}$, and that is equal to one absolute point from the time perspective (when we look at the world from outside), or an isotropic Euclidean space $(c,0)$ as normal complex numbers $\mathbb{C}$, but with one lower dimension, and that is the space perspective (when we look from inside). Infinities and singularities occur when we confuse between these two extreme views; because if the observer is situated inside a spatial dimension it will appear to them continuous and infinite, while it forms only one discrete state in the encompassing outer time. As we shall see in section 5.3, General Relativity is the first approximation for inside observers, but since the Universe is evolving we need to describe it by $\mathbb{H}$, from the time perspective. So GR is correct every instance of time, because the resulting instantaneous space is continuous, but when the outward time flows these instances will form a series of discrete states that should be described by Quantum Field Theory. If we combine these two descriptions properly, we should be able to eliminate GR singularities and QFT infinities.

In other words, the whole homogeneous space forms a single point in the outer time, and our physical Universe is the dynamic combination of these two extreme states, denoted as space-time. This is the same postulated statement that the geometrical points are perpetually and sequentially fluctuating between 0 (for time) and c (for space), and no two points can be in the state of (existence in) space at the same real instance of time, so the points of space come into existence in one chronological sequence, and they can not last in this state for more than one single moment of time, thus they are being perpetually re-created.

Nonetheless, since it is not possible to accelerate a physical object (to make all its geometrical points) to move at the speed of creation c, one alternative way to reach this speed of light, and thus make a new spatial dimension, is to combine the two orthogonal states $(c,0)$ and $(0,c)$, which is the same as matter-antimatter annihilation, and this is a reversible interaction that can be described by the following equation:

$$(c,0) + (0,c) \rightleftarrows (c,c) + (0,0) \qquad (2)$$

In conclusion, we can distinguish three conclusive scenarios for the complex flow of time:

1. In our usual space-time where matter particles are described by (c,v), we are restricted by the normal arrow of time because $t_i < t_r$, as we described above.

2. In the orthogonal arrow of time, when $t_i > t_r$, from our perspective, the vacuum state is described by $(0,c)$, which can be excited to (v,c) that describes the states of anti-matter.

3. When $t_i = t_r$, we get the Euclidean space (c,c), which is equivalent to the initial vacuum state $(c,0)$, but with one higher spatial dimension. This, however, can not happen by means of mechanical motion, but by combining the two orthogonal times, according to equation 2.

Therefore, there are two orthogonal arrows of time $(c,0)$ and $(0,c)$, that can combine and split between the states of $(0,0)$ and (c,c), which all together correspond to the four elemental states, or classical elements, whose quintessence is the Single Monad (Haj Yousef, 2014).

On the other hand, as we can see from Figure 1, the space-time interval can be obtained from: $\|s\| = \sqrt{x^2 + y^2 + z^2 - (ct_i)^2} = \sqrt{r^2 - (ct_i)^2}$, or $\|s\| = \sqrt{x^2 - (ct_i)^2}$ for motion on x-axis only. Alternatively, we can now use the new time-time interval which is the modulus of complex time: $\|t_c\| = \sqrt{t_r^2 - t_i^2}$, and it is indeed the same proper time, τ, in Special Relativity:

$$\begin{aligned}\|t_c\| &= \sqrt{(x/c)^2 - t_i^2} \\ &= t_i\sqrt{(x^2/t_i^2)/c^2 - 1} \\ &= t_i\sqrt{v^2/c^2 - 1} \\ &= -t_i/\gamma = -\tau\end{aligned} \qquad (3)$$

The reason why we are getting the negative signature here is because we exist in the imaginary dimension, and that is why we need some "time extension" to perceive the dimensions of space, that is the real dimension. For example, we need at least three instances to imagine any simple segment; one for each side and one for the relation between them; so we need infinite time to conceive the details of all space. If we exist in the real dimensions of space we would conceive it all at once, as what happens in the event horizon of a black hole. So for us it appears as if time is real and space is imaginary, while the absolute reality is a reflection, and the actual Universe is the dynamic and relative combination of these two extreme states.

This essential property, that the outward time is effectively negative with respect to the real flow of time, will be inherited by the velocity, momentum and even energy; all of which will be similarly negative in relation to their real counter part. It is this fundamental property that will enable the derivation of the relativistic momentum-energy relation, the equivalence between inertial and gravitational masses, in addition to allowing energy and mass to become imaginary, negative and even multidimensional. This will be discussed further in sections 6.1, 6.2 and 6.4, respectively.

5.3 General Relativity Approximation:

The representation of space-time with imaginary time was used in the early formulation of Special Relativity, by Poincare (Poincaré, 1906), and even Minkowski (Einstein, 2010), but because there were no substantial reasons to treat time as imaginary, Minkowski

5 The Dynamic Formation of Dimensions:

had to introduce the four-dimensional space-time: (x,y,z,t), with Lorentzian metric $(+,+,+,-)$, in which time and space are treated equally, except for the minus. This four-dimensional space later became necessary for General Relativity, due to the presence of gravity, which required Riemannian geometry to evaluate space-time curvatures.

In the split-complex hyperbolic geometry, Lorentz transformations become rotations in the imaginary plane (Fjelstad, 1986), and according to the new discrete symmetry of the time-time frame, this transformation will be equally valid between inertial and non-inertial frames alike, because the dynamic relation between the real and imaginary parts of time implies that the instantaneous velocity in the imaginary time is always **zero** (see also Figure 1), whether the object is accelerating or not. Therefore, in addition to Lorentz transformations, this essential characteristic of the dynamic complex-time geometry allows direct derivation of the equivalence principle that lead to General Relativity. This will be discussed further in sections 6, 6.1 and 6.3 below.

In the Theory of Relativity, we need to differentiate between inertial and non-inertial frames, because we are considering the "apparent velocity", since the observer is measuring the change of position (i.e. space coordinates) with respect to time, thus implicitly assuming their real co-existence and continuity; so considering motion to be real transmutation, and that is why space and time are considered continuous and differentiable. The observer is therefore not realizing the fact that the dimensions of space are being sequentially re-created within the inner levels of time, as we described above. This sequential re-creation is what makes space-time complex and granular, in which case the instantaneous velocity is always zero, while the apparent physical velocity is a result of the superposition of all the velocities of the individual geometrical points N, which constitute the object of observation, and each of which is either zero, in the outer time, or c, in the inner time, as can be calculated by equation 5. So in this hidden discrete symmetry of space, motion is a result of re-creation in the new places rather than gradual and infinitesimal transmutation from one place to the other. Moving objects do not leave their places to occupy new adjacent positions, but they are successively re-created in them, so they are always at rest in any position along the path.

When we realize the re-creation of space at the only real speed c, and thus consider the apparent velocity of physical objects to be genuinely imaginary, we will automatically obtain Lorentz transformations, equally for velocity, momentum and energy (which will become also complex, as explained further in sections 6.3 and 6.4), without the need for introducing the principle of invariance of physics laws, so we do not need to differentiate between inertial and non-inertial frames, because the instantaneous velocity is zero in either case. As an extra bonus, we will also be able to derive the mass-energy equivalence relation $E = mc^2$ without introducing any approximation or un-mathematical induction, and this relation is indeed the same equivalence between gravitational and inertial masses. All this is treated in section 6 below.

Therefore, the non-Euclidean Minkowski space-time continuum is the first global approximation of the metaphysical reality (of Oneness, or sequential re-creation from one single point), just as the Euclidean Minkowski space-time is a local approximation when the effect of gravity is neglected, while the Galilean space is the classical approximation for non-relativistic velocities. These three relative approximations are still serving very well in describing the respective physical phenomena, but they can not describe the actually metaphysical reality of the Universe, which is dynamically re-creating the geometry of space-time itself, and what it contains of matter particles. As Hawking had already noticed: "In fact one could take the attitude that quantum theory, and indeed the whole of physics, is really defined in the Euclidean region and that it is simply a consequence of our perception that we interpret it in the Lorentzian regime." (Hawking, 1979). The Duality of Time explains exactly that the source of this deceptive perception is the fact that we do not witness the metaphysical perpetual re-creation process, but, being part of it, we always see the Universe after it is re-created, so we "imagine" that this existence is continuous, and thus describe it with the various laws of Calculus and Differential Geometry, that implicitly suppose the continuity of space and the co-existence of matter particles in space and time.

In other words, normal observers, since they are part of the Universe, are necessarily approximating the reality, at best in terms of non-Euclidean Minkowskian space, and this approximation is enough to describe the macroscopic physical phenomena from the point of view of observers (necessarily) situated inside the Universe. However, this will inevitably lead to singularities at extreme conditions because, being inside the Universe, observers are trying to fit the surrounding infinite spatial dimensions in one instance of time, which would have been possible only if they are moving at the speed of light, or faster, and in this case a new spatial dimension is formed and the Universe would become confined but now observed from a higher dimension.

For example, we normally see the Earth flat and infinite when we are confined to a small region on its surface, but we see it as a finite semi-sphere when we view it from outer space. In this manner, therefore, we always need one higher dimension to describe the (deceptive, and apparently infinite) physical reality, in order to contain the curvatures (whether they are intrinsic or extrinsic), and that is why Riemannian geometry is needed to describe General Relativity.

Therefore, since using higher dimensions to describe the reality behind physical existence will always lead to space-time singularities, the Duality of Time Theory is working with this same logic, but backward, by penetrating inside the dimensions of space, as they are dynamically formed in the inner levels of time, down to the origin that is the zero-dimensional metaphysical point, which is the unit of space-time geometry. The Duality of Time Theory is therefore penetrating beyond the apparently-continuous physical existence, into its instantaneous or perpetual dynamic formation through the real flow of time, whose individual discrete instances can accommodate only one metaphysical or geometrical point at a time, that then correlate, or entangle, into physical objects that are kinetically evolving in the normal level of time that we encounter.

At the level of this (unreal) physical multiplicity, any attempt to quantize space-time is destined to fail, because we always need a predefined background geometry, or topology, to accommodate multiplicity and define the respective relations between its various entities. In contrast, the background geometry of the Duality of Time Theory is "void", which is an absolute mathematical vacuum that has no structure or reality, while also explaining how the physical vacuum is dynamically formed by simple chronological recurrence. So, apart from natural counting, the Duality of Time does not rely on any predefined geometrical structure, but it explains how space-time geometry itself is re-created as dynamic and genuinely-complex structure.

The fact that each frame of the inner time (which constitutes space) appears as one instance on the outward time is what justifies treating time as imaginary with relation to space, thus orthogonal on it. In this dynamic creation of space in the complex time, the outward time is discrete and imaginary, while space becomes continuous with relation to this outer time, but this is only relative to the dimension in which the observer is situated, so for example: the 2D plane is itself continuous with relation to its inner dimensions but it forms one discrete instance with relation to the flow of time in the encompassing 3D, which then appears internally continuous but discrete with regard to the encompassing outward time. For this reason perhaps, although representing Minkowski space-time in terms of Clifford geometric algebra $G4 = G(M^4)$ employing bivectors (Pavšič, 2005), or even the spinors of complex vector space (Lounesto, 2001), allowed expressing the equations in simple forms, but it could not discover the intrinsic granularity of space-time without any background, since it is still working on the multiplicity level, and not realizing the sequential re-creation process.

5.4 Aether and the Cosmological Constant Problem:

Aether was described by ancient philosophers as a thin transparent material that fills the upper spheres where planets swim. The concept was also used again in the 18th century to explain the propagation of light and gravitation. This continued until the late 19th century in what is called: luminiferous aether, or light-bearing aether, which was needed to allow the wave-based light to propagate through empty space.

The concept of aether was contradictory because this medium must be invisible, infinite and without any interaction with physical objects. Therefore, after the development of Special Relativity, aether theories became scientifically obsolete, although Einstein himself said that his model could itself be thought of as an aether, since empty space now has its own physical properties (Rucker, 2012). In 1951, Dirac reintroduced the concept of aether in an attempt to address the perceived deficiencies in current models (DIRAC, 1951), thus in 1999 one proposed model of dark energy has been named: quintessence (Zlatev et al., 1999), or the fifth fundamental force

(Krasznahorkay et al., 2016). Also, as a scalar field, the quintessence is considered as some form of dark energy which could provide an alternative postulate to explain the observed accelerating rate of the expansion of the Universe, rather than Einstein's original postulate of cosmological constant (Ratra and Peebles, 1988; Caldwell et al., 1998).

The classical concept of aether was rejected because it required ideal properties that could not be attributed to any physical medium that was thought to be filling the otherwise empty space background which was called vacuum. With the new dynamic creation, however, those ideal properties can be explained, because aether is no longer something filling the vacuum, but it is vacuum itself, that is perpetually being re-created at the absolute speed of light. Thus its state is described by: $(c,0)$ as we explained in section 5.1 above, which indicates infinite and inert space that is the ground state of matter particles that are then described by (c,v), whereas the absolutely-empty mathematical space is now called void and its state is $(0,0)$.

As we already explained above, this state of $(c,0)$ corresponds to absolute zero temperature, and it is a perfect super-fluid described by Bose-Einstein statistics because its points are non-interacting and absolutely indistinguishable. When this medium is excited or disturbed, matter particles and objects are created as the various kinds of vortices that appear in this super-fluid, and this is what causes the deformation and curvature of what is otherwise described by homogeneous Euclidean geometry. Therefore, the Duality of Time Theory reconciles the classical view of aether with General Relativity and Quantum Field Theory at the same time, because it is now the ground state of particles that are dynamically generated in time.

In Quantum Field Theory, the vacuum energy density is due to the zero-point energy of quantized fields, which originates from the quantization of the simple harmonic oscillations of a particle with mass. This zero-point energy of the Klein-Gordon field is infinite, but a cutoff at Planck length is enforced, since it is generally believed that General Relativity does not hold for distances smaller than this length: $\ell_P = \sqrt{\hbar G/c^3} = 1.616229(38) \times 10^{-35} m$, which corresponds to Planck energy: $E_P = \sqrt{\frac{\hbar v^5}{G}} \approx 1.2209 \times 10^{19} GeV$. By applying this cutoff we can get: $\rho_{vac} = \frac{E_P^4}{8\pi^2\hbar^3 c^3}$, which gives us: $\rho_{vac} \sim 10^{76} GeV^4$. Comparing this theoretical value with the 1998 observations that produced: $\rho_{vac} \sim 10^{-29} GeV^4$, we find 120 orders of magnitude discrepancy, which is known as the vacuum catastrophe (Gell-Mann et al., 1968; Martel et al., 1998; SAHNI and STAROBINSKY, 2000).

The smallness of the cosmological constant became a critical issue after the development of cosmic inflation in the 1980s, because the different inflationary scenarios are very sensitive to the actual value of ρ_{vac}. Many solutions have been suggested in this regard, as it was reviewed by Weinberg (Weinberg, 1989) and Sanhi (SAHNI and STAROBINSKY, 2000), which include various modifications on either General Relativity or Quantum Filed Theory or the way they are linked together. However, because the difference is so huge, none of these speculations came ever close to solving the puzzle.

The 10^{120} discrepancy is actually many orders of magnitudes larger than the number of all atoms in the Universe, which is called Eddington number $N_{edd} = 10^{80}$ (Kragh, 2003). This indicates that there is something substantially wrong in our understanding of the quantum processes at the sub-atomic level. It would not be strange, therefore, if we postulate that this huge number of atoms, or elementary particles, is not real! Yet since we clearly observe multiplicity in our normal level of time, it remains that they are multiplied in the inner levels of the real flow of time.

According to the Duality of Time postulate, this huge discrepancy in the cosmological constant is diminished, and even eliminated, because the vacuum energy should be calculated from the average of all states, and not their collective summation as it is currently treated in Quantum Field Theory. This means that we should divide the vacuum energy density by the number of modes included in the unit volume. Since we took Planck length as the cutoff, this number is:

$$N = (\frac{2\pi}{\ell_P})^3 = 8\pi^3/(1.616229 \times 10^{-35})^3 \approx 10^{117}. \tag{4}$$

This will reduce the discrepancy between the observed and predicted values of ρ_{vac} from 10^{120} into $\approx 10^3$ only. The remaining small discrepancy could now be explained according to quintessence models, which is already described by the Duality of Time as the

5 The Dynamic Formation of Dimensions: 39

ground state of matter. However, more accurate calculations are needed here because all the current methods are approximate and do not take into account all possible oscillations for all the four fundamental interactions.

5.5 Origin of Mass

It is well established in modern physics that mass is an emergent property, and since the Standard Model relies on gauge and chiral symmetry, the observed non-zero masses of elementary particles require spontaneous symmetry breaking, which suggested the existence of the massive Higgs boson, whose own mass is not explained in the model. This Higgs mechanism is part of the Glashow-Weinberg-Salam theory to unify electromagnetic and weak interactions (Weinberg, 1976; Susskind, 1979). However, as we have already explained in section 5.1 above, the ground state of matter according to the Duality of Time Theory is a perfect super-fluid, where masses of elementary particles could arise from interaction with the physical vacuum, in a manner similar to the gap generation mechanism (Avdeenkov and Zloshchastiev, 2011; Dzhunushaliev and Zloshchastiev, 2013; Zloshchastiev, 2011).

Moreover, the Duality of Time Theory provides an even more fundamental and very simple mechanism for mass generation, in full agreement with the principles of Classical Mechanics, as shown further in section 6.2.3. In general, the fundamental reason for inertial mass is the coupling between the particles that constitute the object, because the binding field enforces specific separations between them, so that when the position of one particle changes, a finite time elapses before other particles move, due to the finite speed of light. This delay is the cause of inertial behavior, and this implies that all massive particles are composed of more sub-particles, and so on until we reach the most fundamental particles which should be massless. This description is fulfilled by the Duality of Time Theory, due to the discrete symmetry of the genuinely-complex time-time geometry as described above.

The Duality of Time Theory is based on the Single Monad Model, so the fundamental reason of the granular geometry is the fact that no two geometrical points can exist at the same real instance of time, so they must be re-created in one chronological sequence. This delay is what causes the inertial mass, so physical objects are dynamically formed by the coupling between at least two geometrical points which produces the entangled dimensions. According to the different degrees of freedom in the resulting spatial dimensions, this entanglement is responsible for the various coupling parameters, including charge and mass, which become necessarily quantized because they are proportional to the number of geometrical nodes constituting each state, starting from one individual point for massless bosons. Nevertheless, some bosons might still appear to have heavy masses (in our outer level of time) because they are confined in their lower dimensions in which they are massless, just as the inertial mass of normal objects is exhibited only when they are moved in the outer level of time.

Consequently, there is a minimum mass gap above the ground state of vacuum $(c, 0)$, which is itself also above the void state $(0, 0)$. This is because each single geometrical node is massless on its own dimensions, while the minimum state above this ground state is composed of two nodes which must have non-zero inertial mass because of the time delay between their sequential creation instances. This important conclusion agrees with Yang-Mills suggestion that the space of intrinsic degrees of freedom of elementary particles depends on the points of space-time. It was already anticipated that proving Yang-Mills conjecture requires the introduction of fundamental new ideas both in physics and in mathematics (Jaffe and Witten, Jaffe and Witten). Accordingly, due to the sequential re-creation in the inner levels of time, the Duality of Time Theory introduced the genuinely-complex time-time geometry which revealed this profound discrete symmetry that makes the vacuum physical and dynamic, with super-fluid properties and discrete mass states.

6 Deriving the Principles of Special and General Relativity:

The famous Michelson-Morley experiment in 1887 proved that light travels with the same speed regardless whether it was moving in the direction of the movement of the Earth or perpendicular to it (MICHELSON and MORLEY, 1991). This unexpected result initiated active research that eventually led to Special Relativity in 1905 (Einstein, 2005). The speed of light in vacuum is then

6 Deriving the Principles of Special and General Relativity:

considered the maximum speed which anything in the Universe can attain. Photons, or massless particles, propagate in vacuum at this terminal speed, regardless of the motion of the source or reference frame of the observer. However, even though this was confirmed by many experiments, there is yet no theoretical or philosophical account that could explain the reasons behind this constancy and invariance of the speed of light.

Logically, there are two cases under which a quantity does not increase or decrease when we add or subtract something from it. Either this quantity in infinite, or it exists in orthogonal dimension. As we have already introduced in sections 2 and 3 above, according to the Duality of Time postulate, both these cases are equivalent and correct for the absolute speed of light in vacuum, because it is the speed of creation which is the only real speed in nature, and it is intrinsically infinite (or indefinite), but the reason why we measure a finite value of it is because of the sequential re-creation process; so individual observers are subject to the time lag during which the dimensions of space are re-created. Moreover, since the normal time is now genuinely imaginary, the velocities of physical objects are always orthogonal onto this real and infinite speed of creation.

As demonstrated in Figure 1 and explained in section 5.1 above, one of the striking conclusions of the sequential re-creation in the inner levels of time is the fact that it conceives of only two primordial states: **vacuum** and **void**, whose spatial and temporal superposition is producing the multiplicity of intermediary states that appear in the cosmos as matter particles. Vacuum is the continuous existence in the inner levels of time, and void is the discrete existence in the outer level. These two super states, which correspond to $(0,0)$ and (c,c) respectively, can be conceived as abstract extreme limits, but only a relative superposition of them can ever be observed or measured as relative events in space-time, or inner-outer time that is the complex-time discussed in section 5.

As the real time flows uniformly in the inner levels, it creates the homogeneous dimensions of vacuum, and whenever it is interrupted or disturbed, it makes a new dimension that appears as a discrete instance on the outer imaginary level which is then described as void, since it does not last for more than one instance, before it is re-created again in a new state that may resemble the previous perished states, which causes the illusion of motion, while in reality it is only a result of successive discrete changes. So the individual geometrical points can either be at rest (in the outer/imaginary time) or at the speed of creation (in the inner/real time), while the apparent limited velocities of physical particles and objects (in the total complex time, which forms the physical space-time dimensions) are the temporal average of this spatial combination that may also dynamically change as they are progressing over the outward ordinary time direction.

Therefore, the Universe is always coming to be, perpetually, in "zero time" (on the outward level), and its geometrical points are sequentially fluctuating between existence and nonexistence (or vacuum and void), which means that the actual instantaneous speed of each point in space can only change from $v_{imaginary} = 0$ to $v_{real} = c$, and vice versa. This instantaneous abrupt change of speed does not contradict the laws of physics, because it is occurring in the inner levels of time before the physical objects are formed. Because they are massless, this fluctuation is the usual process that is encountered by the photons of light on the normal outward level of time, for example when they are absorbed or emitted. Hence, this model of perpetual re-creation is extending this process onto all other massive particles and objects, but on the inner levels of time where each geometrical point is still massless because it is metaphysical, while "space" and "mass" and other physical properties are actually generated from the temporal coupling, or entanglement, of these geometrical points, which is exhibited only on the outward level of time.

Accordingly, the normal limited velocity, of physical particles or objects, is a result of the spatial and temporal superposition of these dual-state velocities of its individual points N, thus:

$$|v| = \frac{\sum_N <0|c>}{N} \Rightarrow 0 \leq |v| \leq c. \tag{5}$$

Individually, each point is massless and it is either at rest or moving at the speed of creation, but collectively they have some non-zero inertial mass m, energy E, and limited total apparent velocity v given by this equation 5 above.

Consequently, there is no gradual motion in the common sense that the object leaves its place to occupy new adjacent places, but it is successively re-created in those new places, i.e. motion occurs as a result of discrete change rather than infinitesimal transmutation, so the observed objects are always at rest in the different positions that they appear in (see also Figure 1). This is the same conclusion of the Moving Arrow argument in Zeno's paradox, which Bertrand Russell described as: "It is never moving, but in some miraculous way the change of position has to occur between the instants, that is to say, not at any time whatever." (Haj Yousef, 2018).

This momentous conclusion means that all frames are effectively at rest in the normal (imaginary) level of time, and there is no difference between inertial and non-inertial frames, thus there is even no need to introduce the second principle of Special Relativity (which says that the laws of physics are invariant between inertial frames) neither the equivalence principle that lead to General Relativity. These two principles, which are necessary to derive Lorentz transformations and Einstein's field equations, are implicit in the Duality of Time postulate and will follow directly from the resulting complex-time geometry as it will be shown in sections 6.1 and 6.3 below. Furthermore, it will be also shown in section 6.2 that this discrete space-time structure that results from the genuinely-complex nature of time is the only way that allows exact mathematical derivation of the mass-energy equivalence relation ($E = mc^2$).

In this manner, the Duality of Time postulate, and the resulting perpetual re-creation in the inner levels of time, can explain at once all the three principles of Special and General Relativity, and transform it into a quantum field theory because it is now based on discrete instances of dynamic space, which is the super-fluid state $(c, 0)$ that is the ground state of matter, while the super-gas state $(0, c)$ is the ground state of anti-matter, which accounts for super-symmetry and matter-antimatter asymmetry as we shall discuss further in section 7.1. The other fundamental forces could also be interpreted in terms of this new space-time geometry, but in lower dimensions: $2D+1$, $1D+1$, and $0D+1$, while gravity is in $3D+1$.

6.1 Lorentz Transformations:

As we noted above, it was originally shown by Poincare (Poincaré, 1906) that by using the mathematical trick of imaginary time, Lorentz transformation becomes a rotation between inertial frames. For example, if the space coordinates of an event in space-time relative to one frame are $X = ct + jx$, then its (primed) coordinates $X' = ct' + jx'$ with respect to another frame, that is moving with uniform velocity $v < c$ with respect to the first frame, are: $X' = Xe^{-j\phi} = [ct + jx][\cosh(\phi) + j\sinh(\phi)]$, where $\tanh(\phi) = \sinh(\phi)/\cosh(\phi) = v/c = \beta$, and since: $\cosh^2(\phi) - \sinh^2(\phi) = j^2 = 1$, then: $\gamma = \cosh(\phi) = 1/\sqrt{1 - v^2/c^2} = 1/\sqrt{1 - \beta^2}$.

In the complex-time frame of the Duality of Time postulate, however, the outer time is the (genuinely) imaginary part, while the real part is the inner time that constitutes space, thus the time coordinates: $T = t_r + jt_i = (x/c) + jt_i$ is used instead of space coordinates: $X = ct + jx$. Therefore, the above rotation equations will still be valid but with time, rather than space, coordinates: $T' = Te^{-j\phi} = [t_r + jct_i][\cosh(\phi) + j\sinh(\phi)]$, and then the speed of creation will be the ground state, or the *rest speed*, so when the apparent imaginary velocity v on the outer time is zero there is still the real speed that is the constant and invariant speed of creation, or in general: $V = c + jv \Longrightarrow \beta = v/c = \tanh(\phi), \gamma = \cosh(\phi) = c/\sqrt{c^2 - v^2}$. See also Figure 1, and also Figure 2.

Using the concept of split-complex time, we can easily derive Lorentz factor $\gamma = 1/\sqrt{1 - v^2/c^2}$ for example by calculating the proper time τ as it can be readily seen from Figure 1, which is replicated in Figure 2 that represents complex velocity, for better clarity, and also because we want to stress the fact that the apparent (imaginary) motion in any direction is in fact interrupting the real motion in the inner time that is re-creating space at the absolute speed of light, so that in the end the *actual velocity* is always smaller than c: $v_a = \sqrt{c^2 - v^2}$, which means that everything in space is intrinsically moving at the absolute speed of light in the real flow of time, and its apparent motion in the outward time is causing an apparent (or imaginary) slow down from this real speed, so that when the object is not moving at all it is actually still moving at the absolute speed of light, which becomes zero in relation to other stationary objects because they are all moving at this same real speed.

Lorentz factor is therefore the ratio of the real velocity c over the actual velocity v_a, which is equal $\cosh\phi$, as demonstrated in Figure

6 Deriving the Principles of Special and General Relativity:

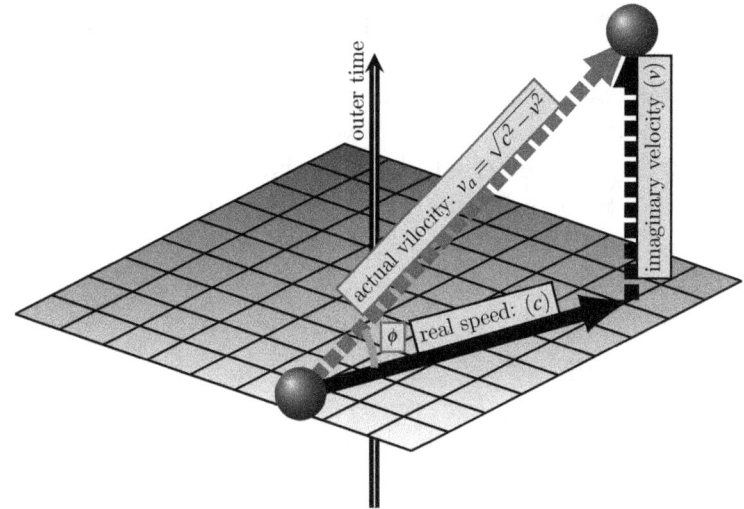

Figure 2: The actual velocity v_a is the modulus of split-complex velocity that combines the real speed $v_r = c$ with the imaginary velocity $v_i = v$, thus: $v_a = \|v_c\| = \|v_r + jv_i\| = \|c + jv\| = \sqrt{c^2 - v^2}$, from which we can easily calculate Lorentz factor as $\gamma = c/v_a = \cosh \phi$.

2:

$$\gamma = c/v_a = \frac{c}{\sqrt{c^2 - v^2}} = 1/\sqrt{1 - v^2/c^2} = \cosh(\phi) \tag{6}$$

6.2 The Mass-Energy Equivalence Relation:

In addition to explaining the constancy and invariance of the speed of light and merging it with the second and third principles of Relativity, the Duality of Time postulate is the only way to explain the equivalence and transmutability between mass and energy ($E = mc^2$). Einstein gave various heuristic arguments for this relation without ever being able to prove it in any theoretical way (Hecht, 2011), neither did anyone else. Based on Doppler effect and Maxwell's theory of radiation, the reasoning that he gave in his 1905 original derivation (Einstein, 1905) was questioned by Planck (Planck, 1906) and shown to be faulty (Ives, 1952). In 1907, Einstein acknowledged the controversy over his derivation (Einstein, 1907), and in the following forty years, he produced more than half dozen proofs that all suffer from unjustified assumptions or approximations. He never succeeded in producing a valid general proof (Ohanian, 2009). In 1955 he wrote in a letter to Carl Seelig: "I had already previously found that Maxwell's theory did not account for the micro-structure of radiation and could therefore have no general validity." (Capria, 2005). In 1990, Rohrlich tried to apply the relativistic Doppler shift, but he also had to introduce various approximations in order to reach the final equation (Rohrlich, 1990). Various other attempts have been made in this regard but until now there is no exact derivation of this famous formula.

It can be readily seen from Figure 3 that the transmutability between mass and energy can only occur in the inner levels of time, because it must involve motion at the speed of light that appears on the normal level of time as instantaneous, hence the same Relativity laws become inapplicable since they prohibit massive particles from moving at the speed of light, in which case $\gamma = 1/\sqrt{1-v^2/c^2} = 1/\sqrt{1-c^2/c^2} = \infty$, so the mass $m = \gamma m_0$ would be infinite and also the energy. In the inner levels of time, however, this would be the normal behavior because the geometrical points are still massless, and their continuous coupling and decoupling is what generates mass and energy on the inner and outer levels of time, respectively, as explained further in section 6.2.3 below.

As we introduced in section 5.1 above, the normal limited velocities of massive physical particles and objects are a result of the spatial and temporal superposition of the various dual-state velocities of their individual points. This superposition occurs in the inner levels of time, where individually each point is massless and it is either at rest or moving at the speed of creation, but collectively they have some non-zero inertial mass m, energy E, and limited total apparent velocity v, which can be calculated from equation 5. We also explained in section 5.3 above, that when we consider this imaginary velocity as being real, the Duality of Time Theory reduces to General Relativity, but when we consider its imaginary character we will uncover the hidden discrete space-time symmetry and we will automatically obtain Lorentz transformation, without introducing the principle of invariance of physics laws. For the same reason, we can see here that the mass-energy equivalence $E = mc^2$ can only be derived based on this profound discreteness that is manifested in dual-state velocity, which then allows the square integration in

6 Deriving the Principles of Special and General Relativity:

Figure 3, because the change in speed is occurring abruptly from zero to c. Otherwise, when we consider v to be real continuous in time, we will get the gradual change which produces the triangular integration with the factor of half that gives the normal kinetic energy $E_k = \frac{1}{2}mv^2$.

Based on this metaphysical behavior in the inner levels of time, we will provide in the following various exact derivations of the mass-energy equivalence relation, in its simple and relativistic forms, directly from the classical equation of mechanical work $E = \int_0^x F.dx$. The first two methods, in sections 6.2.2 and 6.2.3, involve integration (or rather: summation) in the inner time when the velocity changes abruptly from zero to c, or when the mass is generated (from zero to m) in the inner time. This is obviously not allowed on the normal level of time when dealing with physical objects. The third method, in section 6.2.7, gives the total relativistic energy $E = m_0c^2 + \frac{1}{2}mv^2$, by integrating over the inner and outer levels together, while in section 6.2.8 we will derive the relativistic energy-momentum relation directly from the definition of momentum as $p = mv$, also by integrating over the inner and outer levels together and accounting for what happens in each stage. Furthermore, we will see in sections 6.3 and 6.4 that the absolute invariance, and not just covariance, of complex momentum and energy, provide yet other direct derivations because they also lead to $m = \gamma m_0$, that is equivalent with $E = m_0c^2$ or $E = \sqrt{(m_0c^2)^2 + (pc)^2}$ as demonstrated in 1.

Actually, since we have shown previously that the new vacuum $(c,0)$ is a perfect super-fluid, the mass-energy equivalence relation can be easily derived from the equation of wave propagation in such perfect medium: $(c^2 = \partial p/\partial \rho = \partial[(E/m)\rho]/\partial \rho = E/m)$, but we will not discuss that further in this article.

6.2.1 The classical kinetic energy (in the normal time):

In normal classical mechanics, the kinetic energy is the work done in accelerating a particle during the infinitesimal time interval dt, and it is given by the dot product of force F and displacement dx:

$$\begin{aligned}E &= \int_0^x F.dx \\ &= \int_0^t F.vdt \\ &= \int_0^t \frac{d(mv)}{dt}.vdt \\ &= \int v.d(mv),\end{aligned} \qquad (7)$$

thus:

$$E = \int (v^2 dm + mvdv). \qquad (8)$$

Now if we assume mass to be constant, so that: $dm = 0$ (and we will discuss relativistic mass further in section 6.2.4 below), we will get:

$$E = \int (v^2 dm + mvdv) = 0 + m\int_0^v vdv. \qquad (9)$$

So in the classical view of apparently continuous existence, when we consider both space and time to be real, i.e. when we consider an infinitesimally continuous and smooth change in speed from zero to v, the result of this integration will give the standard equation that describes the kinetic energy of massive particles or objects moving in the normal level of time:

$$E_k = \frac{1}{2}mv^2. \qquad (10)$$

The reason why we are getting the factor of "half" in this equation is because the velocity increases gradually with time, which makes the integration equals the area of the triangle as demonstrated by the first arrow in Figure 3.

6.2.2 Method I (abrupt change of speed in the inner time):

The relativistic energy-momentum relation is derived in section 6.2.8 further below, but the simple mass-energy equivalence relation:

6 Deriving the Principles of Special and General Relativity:

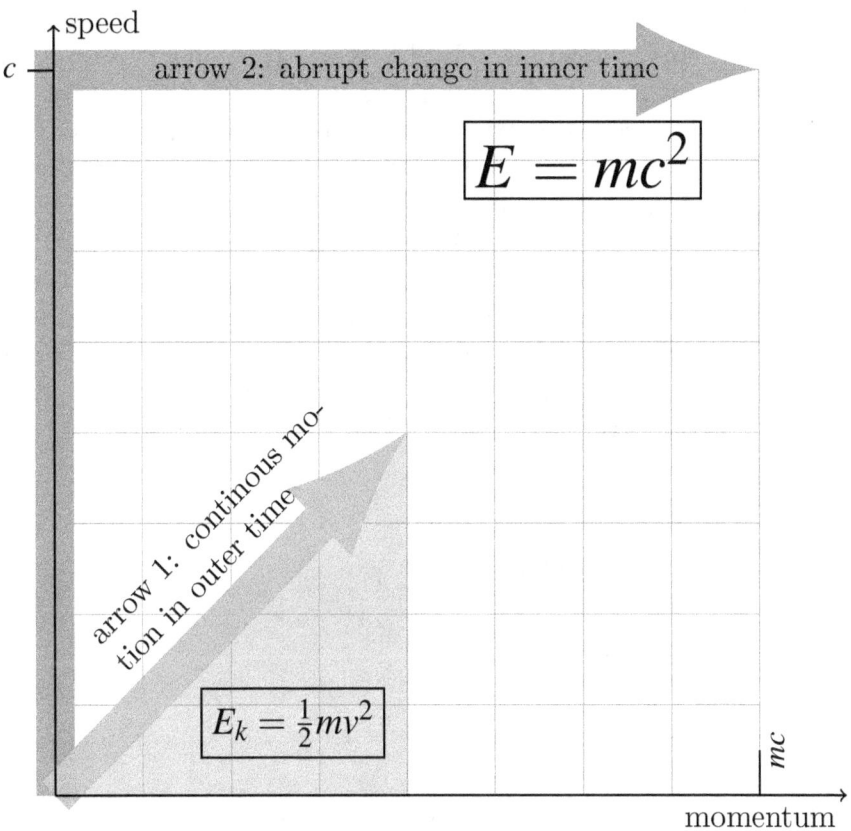

Figure 3: Arrow 1: gradual change of speed in the outer level of time, leading to the classical equation of kinetic energy $E_k = \frac{1}{2}mv^2$. Arrow 2: abrupt change from zero to c in the inward levels of time, leading to the mass-energy equivalence relation $E = mc^2$, which can not happen on the normal level of time because physical objects may not move at the absolute speed of light.

$E = mc^2$ (without the "half") can now be easily obtained from the same integration in equation 9 if, **and only if**, we suppose that the object, whose mass is m, moves from rest to c, or vice versa, in "zero time", which of course will contradict the laws of physical motion because the acceleration would be infinite, and hence the force and the energy. Light does in fact behave in this manner, for example: in pair production, or when emitted or absorbed; but the photon is massless, unlike other particles and objects which have mass and suffer inertia and acceleration.

By introducing the duality of time and the resulting perpetual re-creation, this problem is solved because the conversion between mass and energy takes place, sequentially, in the inner levels of time, on all the massless geometrical points that constitute the particle, and this whole process appears as one instance in the outer level, as demonstrated in Figure 1 above.

So by integrating equation 9 directly from zero to c, which then becomes summation because it is an abrupt change, with only the two states of void and vacuum, corresponding to zero and c, respectively, and since the change in the outward time is zero, and here we also consider $dm = 0$, since the apparent velocity does not change in this case, but we will also discuss relativistic mass in section 6.2.4 below; thus we obtain:

$$\begin{aligned} E &= \int (v^2 dm + mv\,dv) \\ &= 0 + m \int_0^c v.dv \\ &= m \sum_0^c v.dv = mc^2. \end{aligned} \qquad (11)$$

The difference between the above two cases that result in equations 10 and 11 is demonstrated in Figure 3, where in the first case the integration that gives the kinetic energy E_k is the area of the **triangle** below the gradual arrow (1), while in the second case it is the area of the **rectangle** below the right angle arrow (2).

6.2.3 Method II (generating mass in the inner time):

We explained in section 5.5 above how the Duality of Time Theory provides a fundamental mass generation mechanism in addition to its super-fluid vacuum state where mass can be generated via the interaction with this physical vacuum. Hence we can also arrive to the mass-energy equivalence relation directly from the starting equation 8 in an alternative manner if we consider a sudden decoupling, or disentanglement, of the geometrical points that couple together in order to constitute the physical particle that appears in the outer level of time with inertial mass m moving at an apparent (imaginary) velocity v, so when these geometrical points are disentangled to

6 Deriving the Principles of Special and General Relativity: 51

remain at their real speed (of light), the mass m converts back into energy E while the apparent velocity does not change, because this process is happening in the inward levels of time which appears outwardly as instantaneous. Thus, if we put $dv = 0$ in equation 8 and integrate over mass from m to zero (where $v = c$), or vice versa, we get:

$$E = \int (v^2 dm + mv\,dv)$$
$$= c^2 \int_0^m dm + mv(0) = mc^2. \tag{12}$$

Unlike the classical case in equation 10 where the change in speed occurs in the normal outward level of time, these simple derivations (in equations 11 and 12) would not have been possible without considering the inner levels of time, which appears outwardly as instantaneous.

6.2.4 Note I (relativistic mass):

If we want to consider mass to be variable with speed as in early Special Relativity, and distinguish between rest mass: m_0 and relativistic mass: m, according to the standard equation that uses Lorenz factor: $m = \gamma m_0 = m_0 / \sqrt{1 - v^2/c^2}$, then we can arrive to the equation $E = m_0 c^2$ by calculating the derivative dm/dv which in this case will not be equal to zero as we required in equation 9 above. However, the above relativistic equation of mass ($m = \gamma m_0$) is only obtained based on the same mass-energy equivalence relation that we are trying to prove here, so in this case it will be a circular argument. Therefore, the two equations: $E = m_0 c^2$ and $m = \gamma m_0$ are equivalent, and deriving one of them will lead to the other. See also 1 for how to derive the $E = m_0 c^2$ from $m = \gamma m_0$.

6.2.5 Note II (the mass-energy duality):

We can conclude, therefore, that on the highest existential level, there is either energy in the form of massless active waves moving at the speed of creation in the inner levels of time, or passive mass in the form of matter particles that are always instantaneously at rest in the outer level of time, not the two together; that is what happens in the real flow of time. The various states of massive objects and particles, as well as thermal radiations and energy, are some spatial and temporal superposition of these two primordial states of their metaphysical constituents, so some particles will be heavier than others, and some will have more kinetic energy. In any closed system, such as an isolated particle, atom, or even objects, the contributions to this superposition state come from all the states in the system that are always fluctuating between mass and energy, or void and vacuum, corresponding to $v_i = 0$ and $v_r = c$, respectively, so on average the total state is indeterminate, or determined only as a probability distribution, as far as it is not detected or measured. This wave-particle duality will be discussed further in section 7.

Consequently, everything in the Universe is always fluctuating between particle state and wave state, or mass and energy, which can be appropriately written as: $m_0 c^2 \rightleftharpoons hf$. This means that a particle at rest with mass m_0 can be excited into a wave with frequency f, and the opposite is correct when the wave collapses into particle. Either zero mass at the speed of creation, or (instantaneously) zero energy at rest, or: either energy in the active existence state or mass in the passive nonexistence state. The two cannot happen together on the primary level of time, but a mixture or superposition of various points is what causes all the phenomena of motion and interaction between particles with limited velocities and energy on the outward level of time.

6.2.6 Note III (the effective mass):

Therefore, even when the object is moving at any velocity that could be very close to c, its instantaneous velocity is always zero: $v_i = 0$, at the actual time of measurement, and its mass will still be the same rest mass: m_0, because it is only detected as a particle, while its kinetic energy will be given by its relativistic mass: $E_k = \frac{1}{2}mv^2$, and then its total energy equivalence, with relation to an observer moving at a constant (apparent) velocity v, will be given by:

$$E = E_0 + E_k = m_0 c^2 + \frac{1}{2}mv^2. \tag{13}$$

Thus, with the help of Lorentz factor $\gamma = \sqrt{1 - v^2/c^2}$, we could get rid of the confusion between "rest mass" and "relativistic mass" and just call it mass m since the above equation 13 describes energy and not mass:

$$E = E_0 + E_k = mc^2 + \frac{1}{2}\gamma mv^2. \tag{14}$$

Which means that the mass of any particle is always the rest mass, it is not relativistic, but its energy is relativistic, primarily because energy is related to time and motion or velocity. However, since we have been using m_0 all over this article, we will keep using it as the rest mass, and $m = \gamma m_0$ as the effective mass, unless stated otherwise.

6.2.7 Method III (the total relativistic energy):

The total relativistic energy in equation 13 can also be obtained by integrating the starting equation 8 over the inner and outer time together, since in the inner time the rest energy $E_0 = m_0 c^2$, or mass m_0, is generated at the speed of creation c as a result of the instantaneous coupling between the geometrical points which constitute the particle of mass m_0 (thus $dv = c$ and $dm = m_0$ in the inner time), and in the outer time the kinetic energy $E_k = \frac{1}{2}mv^2$ is generated as this mass m_0 moves gradually so its apparent velocity changes by dv, which corresponds to increasing the effective mass from m_0 to m, thus we can integrate:

$$\begin{aligned} E &= \int (v^2 dm + mv\, dv) \\ &= \int_{0,0}^{c,m_0} v^2 dm + \int_{0,m_0}^{v,m} mv\, dv. \end{aligned} \tag{15}$$

Thus we get the same equation 13:

$$\begin{aligned} E &= m_0 c^2 + \frac{1}{2}(m)(v^2) - \frac{1}{2}(m_0)(0) \\ &= m_0 c^2 + \frac{1}{2}mv^2. \end{aligned} \tag{16}$$

This equation can also be given in the general form that relates the relativistic energy and momentum (see 1 for how to convert between these two equations):

$$E = \sqrt{(m_0 c^2)^2 + (\gamma m_0 cv)^2} \\ = \sqrt{(m_0 c^2)^2 + (pc)^2} = c|\vec{P}|; p = \gamma m_0 v. \quad (17)$$

This last equation, that is equivalent to equation 16, will be also derived in section 6.2.8 starting from the definition of momentum as $p = mv$, but because it is genuinely complex, and hyperbolic, the imaginary part of momentum will have negative contribution just as we have seen for the outer time when we discussed the arrow of time in section 5.2 above.

Again, however, a fundamental derivation of this relativistic energy-momentum equation 17 is not possible without the Duality of Time postulate. All the current derivations in the literature rely on the effective mass relation: $m = \gamma m_0$, which is equivalent to the same relation we are trying to derive (see above and also 1), while finding this equation from the four-momentum expression, or space-time symmetry, relies on induction rather than rigorous mathematical formulation.

6.2.8 Method IV (complex momentum and energy-momentum relation)

The equation: $|\vec{P}| = E/c = \sqrt{(m_0 c)^2 + (\gamma m_0 v)^2}$, of the relativistic energy-momentum, can be also derived directly from the fundamental definition of momentum: $p = mv$, when we include the metaphysical creation of mass in the inner levels of time, in addition to its physical motion in the outer level, and by taking into account the complex character of time. Thus we need to integrate $dp_c = d(mv)$ over the inner and outer levels, according to what happens in each stage; first by integrating between $(0,0)$ and (c, m_0) on the inner real levels of time where the particle is created, or being perpetually re-created, at the speed of creation c, and this term makes the real part of the complex momentum $p_r = m_0 c$. Then we integrate between $(0, m_0)$ and (v, m) on the outer imaginary level of time where the particle whose mass is m_0 gains an apparent velocity v, and thus its effective mass increases from m_0 to m, and this term makes the imaginary

6 Deriving the Principles of Special and General Relativity:

part of the complex momentum $p_i = mv = \gamma m_0 v$:

$$
\begin{aligned}
p_c &= \int_{0,0}^{m,v} d(mv) \\
&= \int_{0,0}^{m,v} (vdm + jmdv) \\
&= \int_{0,0}^{c,m_0} vdm + j\int_{0,m_0}^{v,m} mdv.
\end{aligned}
\tag{18}
$$

The first term gives us the real momentum $p_r = \int_{0,0}^{c,m_0} vdm = m_0 c$, while the second term gives the imaginary momentum $p_i = \int_{0,m_0}^{v,m} mdv = 0(m - m_0) + m(v - 0) = mv = \gamma m_0 v$. So the total complex momentum is: $p_c = p_r + jp_i = m_0 c + jmv = m_0 c + j\gamma m_0 v$. Hence the modulus of this total complex momentum is given by:

$$\|p_c\| = \sqrt{(m_0 c)^2 - (\gamma m_0 v)^2}. \tag{19}$$

Again, we notice here that the contribution of the imaginary momentum $p_i = mv = \gamma m_0 v$ is negative with relation to the real momentum $p_r = m_0 c$, just like the case of t_i and v_i as we have seen in sections 5.2 and 6.1, and as it will be also the case for complex energy as we shall see in section 6.4 further below. All this is because the normal time, or physical motion, is interrupting the real creation, which is causing the disturbance and curvature of the otherwise infinite homogeneous Euclidean space that describes the vacuum state $(c, 0)$.

Therefore, to obtain the relativistic energy-momentum relation from equation 19, we simply multiply by c:

$$
\begin{aligned}
E &= c\|\vec{P}\| = c\|p_c\| \\
&= \sqrt{(m_0 c^2)^2 - (\gamma m_0 vc)^2} \\
&= \sqrt{(m_0 c^2)^2 - (pc)^2}; p = \gamma m_0 v.
\end{aligned}
\tag{20}
$$

These equations 19 and 20 above, with the negative sign, do not contradict the equation in current Relativity: $E = \sqrt{(mvc)^2 + (m_0 c^2)^2}$

which treats energy as scalar and do not realize its complex dimensions (see also section 6.4 below). Practically, in any mass-energy interaction or conversion, the negative term will be converted back to positive because when the potential energy m_0c^2 is released, in nuclear interactions for example, this means that it has been released from the inner levels of time where it is captured as mass, into the outer level to become kinetic energy or radiation. In other words: the absorption and emission of energy or radiation, nuclear interactions, or even the acceleration and deceleration of mass, are simply conversions between the inner and outer levels of time, or space and time, respectively. Eight centuries ago, Ibn al-Arabi described this amazing observation by saying: "Space is rigid time, and time is liquid space." (Haj Yousef, 2017).

This derivation of the relativistic energy-momentum relation from the fundamental definition of momentum $p = mv$ is based on the Duality of Time concept, by taking into account the complex nature of time, as hyperbolic numbers, which is why the contribution of the imaginary term appears here as negative in equation 20. As we discussed in section 5.3, when we do not realize the discrete structure of space-time geometry that results from this genuinely-complex nature of time, the Duality of Time Theory is reduced to General Relativity, which considers both space and time to be real, and then we take the apparent rather than the complex velocity whose instantaneous value is always zero; so this negative sign in equations 19 and 20 above will appear positive, as if we are treating space-time to be spherical ($\mathbb{R}^4$) rather than hyperbolic ($\mathbb{H}^4$).

Therefore, when we take into account the complex nature of time as we described in sections 5.1 and 6.1 above (or Figures: 1 and 2), energy and momentum will be also complex and hyperbolic. This significant conclusion, that is a result of the new discrete symmetry, will introduce an essential modification on the relativistic energy-momentum equation which will lead to the derivation of the equivalence principle and allows energy to be imaginary, negative and even multidimensional, as it will be discussed further in sections 6.3 and 6.4.

6.3 The Equivalence Principle of General Relativity

In moving from Special to General Relativity, Einstein observed the equivalence between the gravitational force and the inertial force experienced by an observer in a non-inertial frame of reference. This is roughly the same as the equivalence between active gravitational and passive inertial masses, which has been later accurately tested in many experiments (Schlamminger et al., 2008; Reasenberg et al., 2012), but there is no direct mathematical derivation to this principle apart from the famous spacecraft accelerator thought experiment which relies on induction.

When Einstein combined this equivalence principle with the two principles of Special Relativity, he was able to predict the curved geometry of space-time, which is directly related to its contents of energy and momentum of matter and radiation, through a system of partial differential equations known as Einstein field equations.

We explained in section 6.2 above that an exact derivation of the mass-energy equivalence relation $E = mc^2$ is not possible without postulating the inner levels of time, and that is why there is yet no single exact derivation of this celebrated equation. For the same reason indeed, there is also no mathematical derivation of the equivalence principle that relates gravitation with geometry, because it is actually equivalent to the same relation $E = mc^2$ that reflects the fact that space and matter are always being perpetually re-created in the inner time, i.e. fluctuating between the particle state $(0,0)$ and wave state (c,c), thus causing space-time deformation and curvature.

Due to the discrete structure of the genuinely-complex time-time geometry, as illustrated in Figure 1, the complex momentum p_c should be invariant between inertial and non-inertial frames alike, because effectively all objects are always at rest in the outer level of time, as we explained in section 5 above. This means that complex momentum is always conserved **even when the velocity changes**, i.e. as the object accelerates between non-inertial frames!

This invariance of momentum between non-inertial frames is conceivable, because it means that as the velocity increases (for example), the gain in kinetic momentum $p_i = mv$ (that is the imaginary part) is compensated by the increase in the effective mass: $m = \gamma m_0$ due to acceleration, which causes the real part $p_r = mc$ also to increase, but since $p_c = p_r + jp_i$ is hyperbolic, thus its modulus

$\sqrt{(mc)^2 - (mv)^2}$ remains invariant, and this what makes the geometry of space (manifested here as mc) dynamic, because it must react to balance the change in effective mass. Therefore, a closed system is closed only when we include all its contents of mass and energy (including kinetic and radiation) as well as the background space itself, which is the vacuum state $(c, 0)$, and the momentum of all these constituents is either $p_r = mc$, when they are re-created in the inner levels, or $p_i = mv$ for physical objects moving in the normal level of time. For such a conclusive system, the complex momentum $p_c = p_r + jp_i$ is absolutely invariant.

Actually, without this exotic property of momentum it is not possible at all to obtain an exact derivation of $m = \gamma m_0$ which is equivalent to $E = m_0 c^2$ as we mentioned in section 6.2 above, and also in 1 below. These experimentally verified equations are correct if, **and only if**, the modulus $\|p_c\|$ is always conserved. For example when the object accelerates from zero to v, and then the effective mass changes from m_0 to m, thus we get:

$$\sqrt{(mc)^2 - (mv)^2} = \sqrt{(m_0 c)^2 - (m_0 \times 0)^2} = m_0 c \qquad (21)$$

Since this previous equation 21 is equivalent to: $m = m_0 c / \sqrt{c^2 - v^2} = \gamma m_0$, therefore, in addition to the previous methods in equations 11 and 12, and the relativistic energy-momentum relation in equation 20, the mass-energy equivalence relation: $E = m_0 c^2$ can now be deduced from equation 21 as it is shown in 1 below, because, as we mentioned in section 6.2.4 above, the equations: $E = m_0 c^2$ and $m = \gamma m_0$ are equivalent, and the derivation of one of them leads to the other, while there is no exact derivation of either form in the current formulation of Special or General Relativity.

This absolute conservation of complex momentum under acceleration leads directly to the equivalence between active and passive masses, because it means that the total (complex) force: $F_c = \frac{dp_c}{dt}$ must have two components; one that is related to acceleration as v changes in the outer time t_i, which is the imaginary part, and this causes the acceleration: $F_i = ma = mdv/dt$, so m here is the passive mass, while the other force is related to the change in effective mass $m = \gamma m_0$, or its equivalent energy $E = mc^2$, which is manifested as the deformation of space which is being re-created in

6 Deriving the Principles of Special and General Relativity:

the inner levels of time t_r, and this change or deformation is causing the gravitational force F_r that is associated with the active mass; and these two components must be equivalent so that the total resulting complex momentum remains conserved. Therefore, gravitation is a reaction against the disturbance of space from the ground state of bosonic vacuum $(c, 0)$ to the state of fermionic particles (c, v), the first is associated with the active mass in the real momentum mc, and the second is associated with the passive mass in the imaginary momentum mv.

However, as discussed further in section 7, because of the fractal dimensions of the new complex-time discrete geometry, performing the differentiation of this complex function $p_c = mc + jmv$ requires non-standard analysis because space-time is no longer everywhere differentiable (Nottale and Schneider, 1984). So this will not be pursued in this article.

From this conservation of complex momentum we should be able to find the law of gravitation and the stress-energy-momentum tensor which leads to the field equations of General Relativity. Moreover, since empty space is now described as the dynamic aether, gravitational waves become the longitudinal vibrations in this ideal medium, and the graviton will be simply the moment of time mv, just as photons are the quanta of electromagnetic radiations and they are transverse waves in this vacuum, or the moments of space mc. This means that equivalence principle is essentially between photons and gravitons, or between space and time, while electrons and some other particles could be described as standing waves in the space-time; with complex momentum $mc + jmv$, and the reason why we have three generations of fermions is due to the three dimensions of space. This important conclusion requires further investigation, but we should also notice here that the equivalence principle should apply equally to all fundamental forces and not only to gravity, because it is a property of space-time geometry in all dimensions, and not only the $3D$ where gravity is exhibited, as it is also outlined in another publication (Haj Yousef, 2017).

6.4 Complex Energy

Since it is intimately related to time, energy has to have complex, and even multiple intersecting dimensions in accordance with the dimensions of space and matter which are generated in the inner levels of time before they evolve throughout the outer level. We must notice straightforward, however, that not all these levels of energy are equivalent to mass which is only a property of 3D space. In lower dimensions, energy should rather be associated with the corresponding coupling property, such as the electric and color charges. Therefore, it is expected that negative mass is only possible in 4D spatial dimensions, as it has been already anticipated before (Bonnor, 1989; Petit and D'Agostini, 2014).

It is clear initially that, just like time, velocity and momentum that were discussed above, when we take the complex nature of time into account, the kinetic energy $\frac{1}{2}\gamma m_0 v^2$ in equation 13, or pc in the relativistic energy-momentum equation 17, becomes negative with relation to the potential energy mc^2 stored in mass m. Therefore the energy E in equation 15 becomes complex E_c with real E_r and imaginary E_i parts. The real part E_r represents re-creation through the change in mass dm, and the imaginary part E_i represents the kinetic evolution of this mass in the outer time through the change in the apparent velocity dv:

$$E_c = \int (v^2 dm + jmvdv)$$
$$= \int_{0,0}^{c,m_0} v^2 dm + j \int_{m_0,0}^{m,v} mvdv. \qquad (22)$$

The real part is $E_r = m_0 c^2$ and the imaginary part is $E_i = \gamma m_0 vc = pc$, thus we get:

$$\|E_c\| = \sqrt{(m_0 c^2)^2 - (\gamma m_0 vc)^2}$$
$$= \sqrt{(m_0 c^2)^2 - (pc)^2}; p = \gamma m_0 v. \qquad (23)$$

This negative contribution of the kinetic energy, however, does not falsify the current equations 13 and 17, but it means that the potential energy and the kinetic energy are in different orthogonal levels of time and the conversion of potential energy into kinetic energy is like the conversion from the inner time into the outer time, so when they are in the outer time they are added together as in the previous equations because they become both in the same level of time.

Again, just as it is the case with the absolute conservation of momentum that we have seen in section 6.3 above, energy is also always conserved, even when the apparent velocity v changes, since the instantaneous velocity v_i in the outer level of time is always zero, as we have seen in section 5 and Figure 1 above. As it is the case for momentum, this absolute conservation of energy is conceivable because it means that as the velocity changes, the change in kinetic energy $E_i = mvc$ (that is the imaginary part) is compensated by the change in the effective mass: $m = \gamma m_0$ due to motion, which causes the real part of energy $E_r = mc^2$ also to change accordingly, but since $E_c = E_r + jE_i$ is hyperbolic, thus its modulus $\sqrt{(mc^2)^2 - (mvc)^2}$ remains invariant between all inertial or non-inertial frames.

This means that:

$$\|E_c\| = \sqrt{(mc^2)^2 - (mvc)^2} = m_0 c^2. \tag{24}$$

This equation provides even an additional method to derive the mass-energy equivalence, because the left side in this equation can be reduced to mc^2/γ:

$$\begin{aligned} \sqrt{(mc^2)^2 - (mvc)^2} &= mc\sqrt{c^2 - v^2} \\ &= mc^2 \frac{\sqrt{c^2 - v^2}}{c} \\ &= mc^2/\gamma \Rightarrow m = \gamma m_0. \end{aligned} \tag{25}$$

So when we combine the two equations: 24 and 25, we get the effective mass relation: $m = \gamma m_0$ that is equivalent to $E = m_0 c^2$ as we have seen above and in 1.

7 Fractal Space-Time and Quantum Phenomena:

Soon after the discovery of fractals, fractal structures of space-time were suggested in 1983 (Ord, 1983), as an attempt to find a geometric analogue for relativistic quantum mechanics, in accordance with Feynman's path integral formulation, where the typical infinite number of paths of quantum-mechanical particles are characterized as being non-differentiable and fractal (Abbott and Wise, 1981). This theoretical concept suggests that the structure of space-time itself has a fractal dimension, and not only the large-scale distribution of galaxies as confirmed by some observations (Joyce et al., 2005; Hogg et al., 2005), in addition to the abundance of various fractal structures on all physical and biological levels in nature, as was previously described by Mandelbrot (Mandelbrot, 1983).

Accordingly, some theories were constructed based on fractal space-time, including Causal Dynamical Triangulation (Ambjørn et al., 2005) and Scale Relativity (NOTTALE, 1992), which also share some fundamental characteristics with Loop quantum gravity, that is trying to quantize space-time itself (Rovelli, 2011). Actually, there are many studies that have successfully demonstrated how the principles of quantum mechanics can be derived from the fractal structure of space-time (Nottale and Célérier, 2007; JUMARIE, 2001; Cresson, 2003; Adda and Cresson, 2005; Jumarie, 2007), but there is yet no complete understanding of how the dimensionality of space-time evolved to the current Universe. Some multiverse and eternal-inflation theories exhibit fractality at scales larger than the observable Universe (Linde, 1987), while other theories suggest that space-time dimensionality developed gradually from $2D$ at the Planck scale, to become $4D$ at large galactic scales (Ambjørn et al., 2005; Lauscher and Reuter, 2005; NOTTALE, 1992). Fractality also arises in the non-commutative geometry approach to quantum gravity, which tries to understand how fractal space couples with gravitation, and suggests that time is an emergent property (Connes and Rovelli, 1994).

In this regard, based on the concept of re-creation according to the Duality of Time Theory, the Universe is constantly being re-created from one geometrical point, that is $0D$, from which all the current dimensions of space and matter are re-created in the inner levels of time before they evolve in the outer time. Therefore, the total dimension of the Universe becomes naturally multi-fractal and equals to the dynamic ratio of "inner" to "outer" times, because spatial dimensions alone, as an empty homogeneous space, are complete integers, while fractality arises when this super-fluid vacuum, as described in section 5.1 above, starts oscillating in the outer time, which causes all types of vortices that we denote as elementary particles. So we can see how this notion, of space-time having fractal dimensions, would not have any "genuine" meaning unless both the numerator and denominator of the fraction are both of the same nature of time, and this can only be fulfilled by interpreting the complete dimensions of space as inner levels of time.

In the absolute sense, the ratio of inner to outer times is the same as the speed of light, which only needs to be "normalized" in order to express the fractality of space-time; to become time-time. For example, if the re-creation process, that is occurring in the inner levels of time, is not interrupted in the outer time, i.e. when the outer time is zero, this corresponds to absolute vacuum, that is an isotropic and homogeneous Euclidean space, with complete integer dimensions, that is $3D$ in our normal perception, and it is expected to be $4D$ on large cosmological scales. So the speed of light, in the time-time frame, is a unit-less constant that is equal to the number of dimensions that are ideally 3 for a perfect three-dimensional vacuum, which corresponds to the state of super-energy (c, c), as described in section 5.1, but it may condense down to 0 for void, which is absolute darkness that is the super-mass state $(0, 0)$.

The standard value of the speed of light in vacuum is now considered a universal physical constant, and its exact value is $299,792,458$ meters per second. Since 1983, the length of the Meter has been defined from this constant, as well as the international standard for time. However, this experimentally measured value corresponds to the speed of light in actual vacuum that is in fact not exactly empty. The true speed that should be considered as the invariant Speed of Creation is the speed of light in absolute "void" rather than "vacuum", which still has some energy that may interact with the photons, and delay them, but void is real "nothing". Of course, even high vacuum is very hard to achieve in labs, so void is absolutely impossible.

Because we naturally distinguish between space and time, this speed must be measured in terms of meters per second, and it should be therefore exactly equal to $300,000,000 m/s$. The difference between this theoretical value and the standard measured value is what accounts for the quantum foam, in contrast to the absolute void that cannot be excited. Of course all this depends also on the actual definition of the meter, and also the second, which may appear to be conventional, but in fact they are based on the same ancient

Sumerian tradition, included in their sexagesimal system which seems to be fundamentally related to the structure of space-time (Haj Yousef, 2017, Ch. VII).

Therefore, the actual physical dimensions of the (local) Universe are less than three, and they change according to the medium, and they are expected to be more than three in extra-galactic space, to accommodate negative mass and super-symmetry. For example the fractional dimensions of the actual vacuum is simply $3 \times (299,792,458/300,000,000) = 2.99792458$, and the fractional dimensions of water would be $3 \times (225,407,863/300,000,000) = 2.25407863$, and so on for all transparent mediums according to their relative refraction index. Opaque materials could be also treated in the same manner according to their refraction index, but for other light wave-lengths that they may transfer. Dimensionality is a relative and dynamic property, so the Universe is ultimately described by multi-fractal dimensions that change according to the medium, or the inner dimensions (of space), and also wavelength, that is the outer dimension (or time).

7.1 Super-symmetry and its Breaking:

As we noted above, many previous studies have successfully derived the principles of quantum mechanics from the fractality of space-time, but we want in the remaining of this section to outline an alternative description based on the new complex-time geometry. This has been explained with more details in other publications (Haj Yousef, 2014, 2017, 2019, 2018), but a detailed study is required based on the new findings.

As a result of perpetual re-creation, matter in the Universe is alternating between the two primordial states of void and vacuum, which correspond to the two states of super-mass $(0,0)$ and super-energy (c,c), respectively. Since $(0,0)$ is real void or absolute "nothing", it remains only the state of vacuum (c,c), which is a perfectly homogeneous three-dimensional space, according to our normal perception. Therefore, the Universe, as a whole, is in this perfect state of Bose-Einstein Condensation, which is a state of "Oneness", because its geometrical points are indistinguishable and non-interacting, so it is a perfectly symmetrical and homogeneous or isotropic space. Multiplicity appeared out of this Oneness as a result of breaking the

7 Fractal Space-Time and Quantum Phenomena:

symmetry of the real existence, the super-energy (c,c), and its imaginable non-existence, the super-mass $(0,0)$, into the two states of super-fluid $(c,0)$ and super-gas $(0,c)$, which correspond to particles and anti-particles, which are perpetually, and sequentially, annihilating back into energy (c,c) and splitting again, as described by equation 2 above. This process is occurring every moment of time, and this is actually what defines the moments of time, and causes our physical perception and consciousness.

If existence remained in the bosonic (c,c) state, no physical particles will appear, and no "time", since no change or motion can be conceived. Normal (or the outer level of) time starts when the super-fluid state $(c,0)$, which is the aether, is excited into (c,v), which describes physical particles, or fermionic states, while at the same time the orthogonal super-gas state $(0,c)$ is excited into (v,c), which describes anti-particles that are also fermions in their own time, but bosons in our time reference, and vice-versa, because these opposite time arrows are orthogonal, as we described in section 5.2.

7.2 The Exclusion Principle:

Therefore, physical existence happened as a result of splitting this $3D$ ideal space, which introduced the outer level of time in which fermions started to move and take various different (discrete) states. The fundamental reason behind the quantum behavior, or why these states are discrete, is because no two particles, or fermionic states, can exist simultaneously in the outer time, which is the very fact that caused them to become multiple and make the physical matter, so their re-creation must be processed sequentially, and this is the ontological reason behind the exclusion principle. Therefore, since all fermions are kinetically moving in the outer time, which is imaginary, they must exist in different states, because we are observing them from orthogonal time direction, otherwise we would not see them multiple and in various dimensions. In contrast to that, because bosons are in the real level of time with respect to the observer, they all appear in the same state even though they may be many.

7.3 Uncertainty:

On the other hand, if we suppose the particle is composed of N individual geometrical points, each of which is either in the inner or outer levels of time, so their individual speeds are either zero or c, but collectively appear to be moving at the limited apparent velocity v, that can be calculated from equation 5; thus the particle is described by (c,v). Therefore, because only one point actually exists in the real flow of time, the position of this point is completely undetermined, because its velocity is equal to c, while the rest have been already defined, because they are now in the past, and their velocities had been sequentially and abruptly collapsed from c to zero, after they made their corresponding specific contribution to the total quantum state which defines the position of the particle with relation to the observer.

When the number N is very large, as it is the case with large objects and heavy particles, the uncertainty will be very small, because only one point is completely uncertain at the real instance of time. But for small particles, such as the electron, the uncertainty could become considerably large, because it is inversely proportional to N: $\delta x \propto 1/N$. This uncertainty in position will also increase with (the imaginary) velocity v, or momentum $p_i = mv$, because higher physical velocity means that on average more and more points are becoming in the real speed c, rather than rest, as can also inferred from equation 5.

7.4 Collapse of Wave-Function:

Moreover, we can now give an exact account of the collapse of the wave function, since the superposition state of a system of N individual points comes from averaging their dual-state of zero or c, all of which had already made its contribution except the real current one at the very real instance of time of measurement, which is going to be determined right in the following instance. Therefore, because the state of any individual point automatically collapses into zero after it makes its contribution to the total quantum state, once the moment passes, all states are determined automatically, although their eigenstate may remain unknown, as far as it is not measured.

So, as in the original Copenhagen interpretation, the act of measurement only provides knowledge of the state. However, if the number of points in a system is very small, and since the observer is necessarily part of the system, the observation may have a large impact on determining the final eigenstate.

7.5 Schroedinger's Cat:

Accordingly, the state of Schroedinger's cat, after the box is closed, is either dead or alive; so it is already determined, but we only know that after we open the box, provided that the consciousness of the observer did not interfere during the measurement. Any kind of measurement or detection, necessarily means that the observer, or the measuring device, at this particular instance of measurement, is the subject that is acting on the system; and since there is only one state of vacuum and one state of void, at this real instance of time, the system must necessarily collapse into the passive state, i.e. it becomes the object or particle, because at this particular instance of time the observer is taking on the active state. Of course, this collapsing is not fatal, otherwise particles and objects will disappear forever, but they are re-created or excited again into a new state right after this instantaneous collapse, at which time the observer now would have moved back into an indeterminate state, and becomes an object amongst other objects.

7.6 Entanglement and Non-locality:

The uncertainty and non-locality of quantum mechanical phenomena result from the process of sequential re-creation, or the recurrence of only one geometrical point, which is flowing either in the inward or outward levels of time, which respectively produce the normal spatial entanglement as well as the temporal entanglement. Therefore, entanglement is the general underlying principle that connects all parts of the Universe in space and as well as in time, but it is mostly reduced into simple coherence, which may also dissipate quickly as soon as the system becomes complex. In other words: spatial and temporal entanglement is what defines space-time structure, rather than direct proximity. In this deeper sense, the speed of light is never surpassed even in extreme cases, such as the EPR and quantum tunneling, since there is no transmutation, but the object is re-created in new places which could be at the other end of the Universe, and even in a delayed future time.

Consequently, whether the two particles are separated in space or in time, they can still interfere with each other in the same way because they are described by the same wave function either as one single entangled state or two coherent states. In this way we can explain normal as well as single particle interference, since the wave behavior of particles in each case is a result of the instantaneous uncertainty in determining their final physical properties, such as position or momentum, as they are sequentially re-created.

Spatial entanglement occurs between the points in the internal level of time, while temporal entanglement is between the points of the outer level, so in reality it is all temporal since all the points of space and time are generated in one chronological order that first spreads spatially in the inner metaphysical level and then temporally in the outer physical level.

7.7 Causality:

On the other hand, since the whole Universe is self-contained in space, all changes in it are necessarily internal changes only, because it is a closed system. Therefore, any change in any part of the Universe will inevitably cause other synchronizing change(s) in other parts. In normal cases the effect of the ongoing process of cosmic re-creation is not noticeable because of the many possible changes that could happen in any part of the complex system, and the corresponding distraction of our limited means of attention and perception. This means that causality is no more directly related to space or even time; because the re-creation allows non-local and even non-temporal causal interactions.

In regular macroscopic situations, the perturbation causes gradual or smooth, but still discrete, motion or change; because of the vast number of neighboring individual points, so the effect of any perturbation will be limited to adjacent points, and will dissipate very quickly after short distance, when energy is consumed. This kind of apparent motion is limited by the speed of light, because the change can appear infinitesimally continuous in space.

In the special case when a small closed system is isolated as a small part of the Universe, and this isolation is not necessarily spatial isolation, as it is the case of the two entangled particles in the EPR, then the effect of any perturbation will appear instantaneous because it will be transferred only through a small number of points, irrespective of their positions in space, or even in time.

8 Conclusion

The Duality of Time Theory exposes a deeper understanding of time, that reveals the discrete symmetry of space-time geometry, according to which the dimensions of space are dynamically being re-created in one chronological sequence at every instance of the outer level of time that we encounter. In this hidden discrete symmetry, motion is a result of re-creation in the new places rather than gradual and infinitesimal transmutation from one place to the other. When we approximate this discrete motion in terms of the apparent (average) velocity, this theory will reduce to General Relativity.

We have shown that the resulting space-time is dynamic, granular, self-contained without any background, genuinely-complex and fractal, which are the key features needed to accommodate quantum and relativistic phenomena. Accordingly, many major problems in physics and cosmology can be easily solved, including the arrow of time, non-locality, homogeneity, dark energy, matter-antimatter asymmetry and super-symmetry, in addition to providing the onto-logical reason behind the constancy and invariance of the speed of light, that is currently considered an axiom.

We have demonstrated, by simple mathematical formulation, how all the principles of Special and General Relativity can be derived from the Duality of Time postulate, in addition to exact mathematical derivation of the mass-energy equivalence relation, directly from the principles of Classical Mechanics, as well deriving the equivalence principle that lead to General Relativity.

Previous studies have already demonstrated how the principles of Quantum Mechanics can be derived from the fractal structure of space-time, but we have also provided realistic explanation of quantum behavior, such as the wave-particle duality, the exclusion principle, uncertainty, the effect of observers and the collapse of wave function. We also showed that, in addition to being a perfect super-fluid, the resulting dynamic quintessence could reduce the cosmological constant discrepancy by at least 117 orders of magnitude.

1 Appendix: Deducing $E = mc^2$ from $m = \gamma m_0$:

Starting from equation 8 above:

$$E = \int (v^2 dm + mv \, dv), \tag{26}$$

and we can find dm by differentiating $m = \gamma m_0 = m_0/\sqrt{1 - v^2/c^2}$, with respect to dv:

$$\frac{dm}{dv} = \frac{m_0 v/c^2}{\sqrt[3/2]{1 - v^2/c^2}} = \frac{mv/c^2}{1 - v^2/c^2} = \frac{mv}{c^2 - v^2}. \tag{27}$$

From this equation we find: $mv \, dv = c^2 dm - v^2 dm$, and by replacing in equation 26 we get:

$$E = \int (v^2 dm + c^2 dm - v^2 dm) = \int c^2 dm = mc^2. \tag{28}$$

This method, however, can not be considered a mathematical validation of the mass-energy equivalence relation $E = mc^2$, because the starting equation $m = \gamma m_0$ is not derived by any other fundamental method from current Relativity, other than being analogous to the equations of time dilation and length contraction: $t = \gamma t_0$, $L = L_0/\gamma$.

Using the same equation $m = \gamma m_0$ with $E = mc^2$; thus: $E = m_0 c^2 / \sqrt{1 - v^2/c^2}$, we can also derive the relativistic energy-momentum relation, by squaring and applying some modifications:

$$E^2 = \frac{m_0^2 c^4}{1 - v^2/c^2} = \frac{m_0^2 c^4}{1 - v^2/c^2} + \frac{m_0^2 c^2 v^2}{1 - v^2/c^2} - \frac{m_0^2 c^2 v^2}{1 - v^2/c^2}. \tag{29}$$

From this equation we get: $E^2 = c^2 \frac{m_0^2 v^2}{1 - v^2/c^2} + \frac{m_0^2 c^4 - m_0^2 c^2 v^2}{1 - v^2/c^2}$, thus $E^2 = p^2 c^2 + m_0^2 c^2 \frac{c^2 - v^2}{1 - v^2/c^2} = (pc)^2 + (m_0 c^2)^2$, or:

$$E = \sqrt{(pc)^2 + (m_0 c^2)^2} = c\sqrt{(mv)^2 + (m_0 c)^2} = c|\vec{P}|. \tag{30}$$

Again, since this derivation relies originally on the equation: $m = \gamma m_0$, it can not be considered a mathematical validation of the mass-energy equivalence relation.

Author Biography

Sheikh Ramadhan Subhi Deeb
Date of Birth: 27 - May - 1920
Naqshbandi Master at Sheikh Ahmad Kuftaro Foundation in Damascus

Mohamed Haj Yousef
Date of Birth: 17 - June - 1967
http://mhajyousef.ibnalarabi.com
https://www.amazon.com/Mohamed-Haj-Yousef/e/B001JS2LC2
https://www.facebook.com/mhajyousef
http://twitter.com/mhajyousef
Email: mhajyousef@hotmail.com

Mohamed Haj Yousef is a writer and researcher interested in physics, cosmology, philosophy and Islamic thought, especially with regard to mysticism and Ibn al-Arabi. He did his undergraduate studies in Syria where he earned the B.Sc. degree in Solid State Physics from the University of Aleppo in 1989 and a Postgraduate Diploma in Electronics from the same university in 1990. After that, he obtained the Master's degree in Microelectronic Engineering and Semiconductor Physics from the University of Cambridge in the UK in 1992. After a period of teaching, he resumed to get the PhD from the University of Exeter in UK in the year 2005, where he studied the concept of time in Ibn al-Arabi's cosmology and its implications on modern physics, which was published in several books and eventually lead to the Duality of Time Theory. This research was supervised by Prof. James W. Morris, as it is continuously inspired by the spiritual guidance of Sheikh Ramadhan Subhi Deeb, the Naqshbandi master at Sheikh Ahmad Kuftaro Foundation in Damascus.

The author has also published numerous articles in Arabic and English that combines science, philosophy and Islamic thought. Most of these articles are accessible online at: http://www.ibnalarabi.com. He also published several books on the subject of time, and other related subjects in Islamic thought and Sufi mysticism, including:

The Sufi Interpretation of Joseph Story: (The Path of the Heart)
By Mohamed Haj Yousef
Publisher: al-Marifa (Aleppo, Beirut)
Publisher: CreateSpace (Charleston)
Paperback: 410 pages
ISBN-13: 978-1482022445
ISBN-10: 1482022443
First Published: 1999

The Sun from the West: Biography of Ibn al-Arabi and His Doctrine
By Mohamed Haj Yousef
Publisher: Fussilat (Aleppo)
Publisher: CreateSpace (Charleston)
Paperback: 708 pages
ISBN-13: 978-1482020229
ISBN-10: 148202022X
First Published: 2006

1 Appendix: Deducing $E = mc^2$ from $m = \gamma m_0$:

Ibn Arabi - Time and Cosmology
By Mohamed Haj Yousef
Publisher: Routledge (New York, London)
hardback/Paperback: 256 pages
ISBNs:
(paperback) 978-0415664011/0415664012
(hardback) 978-0415444996/0415444993
(electronic) 978-0203938249
First Published: 2007

That Is All Indeed: what the seeker needs
By Mohamed Haj Yousef
Publisher: CreateSpace (Charleston)
Paperback: 74 pages
ISBN-13: 978-1482077421
ISBN-10: 1482077426
First Published: 2010

The Meccan Revelations: (introduction)
By Mohamed Haj Yousef
Publisher: Amazon - kindle
Paperback: 180 pages
ASIN: B00B0G1S5Y
First Published: 2012

The Meccan Revelations: (volume 1 of 37)
By Muhyiddin Ibn Arabi
Trns. by: Mohamed Haj Yousef
Publisher: CreateSpace (Charleston)
Paperback: 400 pages
ISBN-13: 978-1549641893
ISBN-10: 1549641891
First Published: 2012

Ibnu'l-Arabi Zaman ve Kozmoloji
By Mohamed Haj Yousef
(Turkish translation of: Ibn Arabi-Time and Cosmology)
Trns. by: Kadir Filiz
Publisher: Nefes Yayincilik (Istanbul)
Paperback: 256 pages
ISBN-13: 978-6055902377
ISBN-10: 6055902370
First Published: 2013

Biography of Sheikh Ramadan Deeb
By Mohamed Haj Yousef
Publisher: Tayba-al-Garraa (Damascus)
Publisher: CreateSpace (Charleston)
Paperback: 400 pages
ISBN-13: 978-1482014419
ISBN-10: 1482014416
First Published: 2013

1 Appendix: Deducing $E = mc^2$ from $m = \gamma m_0$:

The Discloser of Desires
(turjuman al-ashwaq)
By Muhyiddin Ibn Arabi
Trns. by: Mohamed Haj Yousef
Publisher: CreateSpace (Charleston)
Paperback: 200 pages
ISBN-13: 978-1499769678
ISBN-10: 1499769679
First Published: 2014

The Days of God
By Mohamed Haj Yousef
(Arabic translation of the Single Monad Model of the Cosmos)
Publisher: CreateSpace (Charleston)
Paperback: 488 pages
ISBN-13: 978-1482022919
ISBN-10: 1482022915
First Published: Jun. 2014

The Single Monad Model of the Cosmos
By Mohamed Haj Yousef
Publisher: CreateSpace (Charleston)
Paperback: 352 pages
ISBN-13: 978-1499779844
ISBN-10: 1499779844
First Published: Jun. 2014

DUALITY OF TIME: Complex-Time Geometry and Perpetual Creation of Space
By Mohamed Haj Yousef
Publisher: CreateSpace (Charleston)
Paperback: 360 pages
ISBN-13: 978-1539579205
ISBN-10: 1539579204
First Published: Jan. 2018

ULTIMATE SYMMETRY: Fractal Complex-Time, Quantum Gravity and the Incorporeal World
By Mohamed Haj Yousef
Publisher: Independently published
Paperback: 323 pages
ISBN-13: 978-1723828690
ISBN-10: 1723828696
First Published: Jan. 2019

TIME CHEST: Particle-Wave Duality from Time Confinement to Space Transcendence
(this book)
By Mohamed Haj Yousef
Publisher: Independently published
Paperback: 220 pages
ISBN-13: 978-1793927156
First Published: Apr. 2019

ETERNITY: ab ante - a poste
By Muhyiddin Ibn al-Arabi
Tr. Mohamed Haj Yousef
Publisher: Independently published
Paperback: 112 pages
ISBN-13: 978-1798666555
First Published: expected Jun. 2019

Bibliography

(1965). 73 - on the theory of superconductivity. In D. T. HAAR (Ed.), *Collected Papers of L.D. Landau*, pp. 546 – 568. Pergamon.

Abbott, L. F. and M. B. Wise (1981). Dimension of a quantum-mechanical path. *American Journal of Physics 49*(1), 37–39.

Adda, F. B. and J. Cresson (2005). Fractional differential equations and the schrödinger equation. *Applied Mathematics and Computation 161*(1), 323 – 345.

Ambjørn, J., J. Jurkiewicz, and R. Loll (2005, Sep). Reconstructing the universe. *Phys. Rev. D 72*, 064014.

Avdeenkov, A. V. and K. G. Zloshchastiev (2011). Quantum bose liquids with logarithmic nonlinearity: self-sustainability and emergence of spatial extent. *Journal of Physics B: Atomic, Molecular and Optical Physics 44*(19), 195303.

Bonnor, W. B. (1989). Negative mass in general relativity. *General Relativity and Gravitation 21*(1143), 1157.

Calcagni, G. (2017). Cosmological constant problem. In *Classical and Quantum Cosmology*, pp. 301–388. Springer.

Caldwell, R. R., R. Dave, and P. J. Steinhardt (1998, Feb). Cosmological imprint of an energy component with general equation of state. *Phys. Rev. Lett. 80*, 1582–1585.

Capria, M. (2005). *Physics Before and After Einstein*, pp. 82. IOS Press.

Connes, A. and C. Rovelli (1994). Von neumann algebra automorphisms and time-thermodynamics relation in generally covariant quantum theories. *Classical and Quantum Gravity 11*(12), 2899.

Cresson, J. (2003). Scale calculus and the schrödinger equation. *Journal of Mathematical Physics 44*(11), 4907–4938.

DeWitt, B. S. (1967, Oct). Quantum theory of gravity. iii. applications of the covariant theory. *Phys. Rev. 162*, 1239–1256.

DIRAC, P. A. M. (1951). Is there an aether? *Nature 168*(4282), 906–907.

Dzhunushaliev, V. and K. G. Zloshchastiev (2013). Singularity-free model of electric charge in physical vacuum: Non-zero spatial extent and mass generation. *Central Eur. J. Phys. 11*, 325–335.

Einstein, A. (1905). Does the inertia of a body depend upon its energy-content. *Ann Phys 18*, 639–641.

Einstein, A. (1907). Über die gültigkeitsgrenze des satzes vom thermodynamischen gleichgewicht und über die möglichkeit einer neuen bestimmung der elementarquanta. *Annalen der Physik 327*(3), 569–572.

Einstein, A. (2005). Zur elektrodynamik bewegter körper [adp 17, 891 (1905)]. *Annalen der Physik 14*(S1), 194–224.

Einstein, A. (2010). *The Principle of Relativity; Original Papers by A. Einstein and H. Minkowski. Translated Into English by M.N. Saha and S.N. Bose; With a*, pp. 70–88. General Books LLC.

Fjelstad, P. (1986). Extending special relativity via the perplex numbers. *American Journal of Physics 54*(5), 416–422.

Gell-Mann, M., R. J. Oakes, and B. Renner (1968, Nov). Behavior of current divergences under $su_3 \times su_3$. *Phys. Rev. 175*, 2195–2199.

Haj Yousef, M. A. (2005, May). *The concept of time in Ibn Arabi's cosmology and its implication for modern physics*. Ph.D. thesis, University of Exeter, Exeter, UK. published by Routledge in 2007 as (Ibn Arabi - Time and Cosmology).

Haj Yousef, M. A. (2007). *Ibn Arabi – Time and Cosmology*. London, New York: Routledge.

Haj Yousef, M. A. (2014). *The Single Monad Model of the Cosmos: Ibn Arabi's Concept of Time and Creation*. Charleston: CreateSpace.

BIBLIOGRAPHY

Haj Yousef, M. A. (2017). *DUALITY OF TIME: Complex-Time Geometry and Perpetual Creation of Space*. The Single Monad Model of The Cosmos, Book 2. Charleston: CreateSpace.

Haj Yousef, M. A. (2018). Zeno's paradoxes and the reality of motion according to ibn al-arabi's single monad model of the cosmos. In S. Mitralexis (Ed.), *Islamic and Christian Philosophies of Time, Vernon Series in Philosophy*, Chapter 7, pp. 147–178. Wilmington, USA: Vernon Press.

Haj Yousef, M. A. (2019). *ULTIMATE SYMMETRY: Fractal Complex-Time, the Incorporeal World and Quantum Gravity*. The Single Monad Model of The Cosmos, Book 3. USA: Amazon Independent Publishing Platform.

Hawking, S. (1998). *A Brief History of Time*, pp. 157. Updated and expanded tenth anniversary edition. Bantam Books.

Hawking, S. W. (1979). Euclidean quantum gravity. In M. Lévy and S. Deser (Eds.), *Recent Developments in Gravitation: Cargèse 1978*, pp. 145–173. Boston, MA: Springer US.

Hecht, E. (2011). How einstein confirmed e 0=mc 2. *American Journal of Physics* 79(6), 591–600.

Hogg, D. W., D. J. Eisenstein, M. R. Blanton, N. A. Bahcall, J. Brinkmann, J. E. Gunn, and D. P. Schneider (2005). Cosmic homogeneity demonstrated with luminous red galaxies. *Astrophys. J. 624*, 54–58.

HUANG, K. (2013). Dark energy and dark matter in a superfluid universe. *International Journal of Modern Physics A 28*(28), 1330049.

HUANG, K., H.-B. LOW, and R.-S. TUNG (2012). Scalar field cosmology ii: Superfluidity, quantum turbulence, and inflation. *International Journal of Modern Physics A 27*(26), 1250154.

Ives, H. E. (1952). Derivation of the mass-energy relation. *JOSA 42*(8), 540–543.

Jaffe, A. and E. Witten. Quantum yang-mills theory, official problem description.

Joyce, M., F. Sylos Labini, A. Gabrielli, M. Montuori, and L. Pietronero (2005, November). Basic properties of galaxy clustering in the light of recent results from the sloan digital sky survey. *aap 443*, 11–16.

JUMARIE, G. (2001). SchrÖdinger equation for quantum fractal space–time of order n via the complex-valued fractional brownian motion. *International Journal of Modern Physics A 16*(31), 5061–5084.

Jumarie, G. (2007). The minkowski's space–time is consistent with differential geometry of fractional order. *Physics Letters A 363*(1), 5 – 11.

Kragh, H. (2003). Magic number: A partial history of the fine-structure constant. *Archive for History of Exact Sciences 57*(5), 395–431.

Krasznahorkay, A. J. et al. (2016). Observation of Anomalous Internal Pair Creation in Be8 : A Possible Indication of a Light, Neutral Boson. *Phys. Rev. Lett. 116*(4), 042501.

Lauscher, O. and M. Reuter (2005, November). Asymptotic Safety in Quantum Einstein Gravity: nonperturbative renormalizability and fractal spacetime structure. *ArXiv High Energy Physics - Theory e-prints*.

Linde, A. D. (1987). Eternally existing self-reproducing inflationary universe. *Physica Scripta 1987*(T15), 169.

Lounesto, P. (2001). *Clifford Algebras and Spinors*. Cambridge Handbooks for Langua. Cambridge University Press.

Mandelbrot, B. (1983). *The Fractal Geometry of Nature*. Einaudi paperbacks. 1997.

Martel, H., P. R. Shapiro, and S. Weinberg (1998). Likely values of the cosmological constant. *The Astrophysical Journal 492*(1), 29.

MICHELSON, A. and E. MORLEY (1991). On the relative motion of the earth and the luminiferous ether. *SPIE milestone series 28*, 450–458.

NOTTALE, L. (1992). The theory of scale relativity. *International Journal of Modern Physics A 07*(20), 4899–4936.

Nottale, L. and M.-N. Célérier (2007). Derivation of the postulates of quantum mechanics from the first principles of scale relativity. *Journal of Physics A: Mathematical and Theoretical 40*(48), 14471.

Nottale, L. and J. Schneider (1984). Fractals and nonstandard analysis. *Journal of Mathematical Physics 25*(5), 1296–1300.

Ohanian, H. C. (2009). Did einstein prove e=mc2? *Studies in History and Philosophy of Science Part B: Studies in History and Philosophy of Modern Physics 40*(2), 167 – 173.

Ord, G. N. (1983). Fractal space-time: a geometric analogue of relativistic quantum mechanics. *Journal of Physics A: Mathematical and General 16*(9), 1869.

Panine, M. and A. Kempf (2016, Apr). Towards spectral geometric methods for euclidean quantum gravity. *Phys. Rev. D 93*, 084033.

Pavšič, M. (2005). Clifford space as a generalization of spacetime: Prospects for qft of point particles and strings. *Foundations of Physics 35*(9), 1617–1642.

Petit, J. P. and G. D'Agostini (2014). Cosmological bimetric model with interacting positive and negative masses and two different speeds of light, in agreement with the observed acceleration of the universe. *Modern Physics Letters A 29*(34), 1450182.

Planck, M. (1906). Das prinzip der relativität und die grundgleichungen der mechanik. *Verh. Deutsch. Phys. Ges. 8*, 136–141.

Poincaré, M. H. (1906). Sur la dynamique de l'électron. *Rendiconti del Circolo Matematico di Palermo (1884-1940) 21*(1), 129–175.

Ratra, B. and P. J. E. Peebles (1988, Jun). Cosmological consequences of a rolling homogeneous scalar field. *Phys. Rev. D 37*, 3406–3427.

Reasenberg, R. D., B. R. Patla, J. D. Phillips, and R. Thapa (2012). Design and characteristics of a wep test in a sounding-rocket payload. *Classical and Quantum Gravity 29*(18), 184013.

Rochon, D. and M. Shapiro (2004). On algebraic properties of bicomplex and hyperbolic numbers. *Anal. Univ. Oradea, fasc. math 11*(71), 110.

Rohrlich, F. (1990). An elementary derivation of e=mc 2. *American Journal of Physics 58*(4), 348–349.

Rovelli, C. (2011). Zakopane lectures on loop gravity. *PoS QGQGS2011*, 003.

Rucker, R. (2012). *Geometry, Relativity and the Fourth Dimension*, pp. 125. Dover Books on Mathematics. Dover Publications.

SAHNI, V. and A. STAROBINSKY (2000). The case for a positive cosmological l-term. *International Journal of Modern Physics D 09*(04), 373–443.

Schlamminger, S., K.-Y. Choi, T. A. Wagner, J. H. Gundlach, and E. G. Adelberger (2008, Jan). Test of the equivalence principle using a rotating torsion balance. *Phys. Rev. Lett. 100*, 041101.

Susskind, L. (1979, Nov). Dynamics of spontaneous symmetry breaking in the weinberg-salam theory. *Phys. Rev. D 20*, 2619–2625.

Weinberg, S. (1976, Feb). Implications of dynamical symmetry breaking. *Phys. Rev. D 13*, 974–996.

Weinberg, S. (1989, Jan). The cosmological constant problem. *Rev. Mod. Phys. 61*, 1–23.

Zlatev, I., L. Wang, and P. J. Steinhardt (1999, Feb). Quintessence, cosmic coincidence, and the cosmological constant. *Phys. Rev. Lett. 82*, 896–899.

Zloshchastiev, K. G. (2011). Spontaneous symmetry breaking and mass generation as built-in phenomena in logarithmic nonlinear quantum theory. *Acta Phys. Polon. B42*, 261–292.

www.ingramcontent.com/pod-product-compliance
Lightning Source LLC
Chambersburg PA
CBHW070812220526
45466CB00002B/648